Leslie,
Great to see
you!
John

A VINEYARD
ODYSSEY

A VINEYARD
ODYSSEY

THE ORGANIC FIGHT TO SAVE WINE
FROM THE RAVAGES OF NATURE

JOHN KIGER

ROWMAN & LITTLEFIELD PUBLISHERS, INC.

Lanham • Boulder • New York • Toronto • Plymouth, UK

Published by Rowman & Littlefield Publishers, Inc.
A wholly owned subsidiary of The Rowman & Littlefield Publishing Group, Inc.
4501 Forbes Boulevard, Suite 200, Lanham, Maryland 20706
www.rowman.com

10 Thornbury Road, Plymouth PL6 7PP, United Kingdom

British Library Cataloguing in Publication Information Available

Library of Congress Cataloging-in-Publication Data

Kiger, John I., 1954-
A vineyard odyssey : the organic fight to save wine from the ravages of nature /
John Kiger.
p. cm.
Includes bibliographical references and index.
ISBN 978-1-4422-2190-1 (cloth : alk. paper) — ISBN 978-1-4422-2191-8
(electronic) 1. Viticulture—California—Sonoma Valley. 2. Vineyards—
California—Sonoma Valley. 3. Viticulturists—California—Sonoma Valley. 4.
Wine and wine making—California—Sonoma Valley. I. Title.
SB387.76.C2K54 2013
634.809794'18—dc23
2013004002

Printed in the United States of America

contents

contents

I

IF I'D ONLY KNOWN
THEN WHAT I KNOW NOW

O idium. Even if you don't know what it is, the very sound of it seems menacing, conjuring up images of evil, disease, and pestilence. Which is entirely appropriate, given its history and potential to wreak havoc on the unsuspecting. Still commonly referred to in Europe as oidium, here in the United States it is more commonly known as grape powdery mildew. Mycologists and plant pathologists call it *Erysiphe necator,* which is the botanical name for the fungus that causes grape powdery mildew. But oidium was the name applied to the disease when it was first discovered in the nineteenth century. The word "oidium," despite the rather ominous image it invokes, derives from the Latin word for a small egg, *oidion,* which aptly describes the small egg-shaped spores by which the fungus reproduces.

Call it what you will, it is the scourge of winegrowers everywhere. In most vineyards, every year brings a new battle to prevent and contain outbreaks of powdery mildew. In some, with only modest effort and expense, powdery mildew is little more than a speed bump on the road to harvest. But for most, powdery mildew is an expensive and ever-present threat.

Powdery mildew is unique among the many pests and diseases that afflict wine grapes and vineyards. No other affliction threatens (nearly) every wine grape vineyard in the world, every year. The consequences of ignoring or not understanding this threat can be catastrophic, as the winegrowers of Europe discovered in the nineteenth century. Likewise, the consequences of simply misjudging the threat of powdery mildew can be traumatic, even for the well-informed and well-intentioned winegrower, as I learned all too well in my own vineyard. But it is hardly alone, as a host of insects, fungi, bacteria, and viruses, along with feathered and furry critters, lurk in the vineyards, all of which are capable of sabotaging a promising vintage right under the nose of the unsuspecting grower.

I knew none of this when I set out to become a winegrower. Well, truth is, I didn't know much of anything about growing wine. I knew I liked wine and fancied myself as something of a connoisseur. So I probably knew a great deal more about wine than the average person, but I knew next to nothing about what it really took to get from a grapevine to a bottle of wine. Deb, my wife and partner in this venture, was skeptical at first, and being a city girl, she was no more familiar than I was with the rigors and hazards of farming or winegrowing.

Sure, we lived in California and visited Wine Country regularly. You know the drill: weekends spent touring vineyards, tasting wines, and talking up the winemakers and tasting-room staff. If nothing else, it endowed me with an appreciation for the beauty of places like Sonoma County and the deep connection between the land, the vines, and the wines that we love so much. This proximity and superficial familiarity no doubt contributed to the idea that a winegrowing future was a realistic possibility.

We came to winegrowing rather late in life. It was to be a third career, of sorts, for me. I was in my mid-forties when we set out to plant our own vineyard, by which time I had endured twenty years of schooling followed by twenty years of corporate life. Of course, none of that education or experience was remotely related to agriculture. However, my background in science, engineering, and the warp-speed, ruthlessly Darwinian world of Silicon Valley high tech did serve me well when I jumped into winegrowing full-time. I thought to myself, how hard could this be? I took it as just another intellectual challenge and something new to master, which was more or less a way of life in the business I had come from.

I will admit up front that this vineyard thing was entirely my idea, and it was not an easy sell, but once Deb bought in, she was all in. It's not always that way. We have a few friends with vineyards or wineries, where one person has the passion and the other wants little or nothing to do with it (apart from living in a nice place and drinking lots of good wine!). In our case, it was meant to be a partnership from day one. Deb is more naturally risk averse and detail oriented, which is a nice compliment to my tendencies to jump into new things and manage more from the big picture than the nitty-gritty details.

The wine business is unusual in that it attracts a great many participants from unrelated walks of life, most with little or no relevant experience beyond the passionate appreciation of fine wines. That certainly described us when we set out to buy a vineyard in quest of a lifestyle centered on growing grapes and making wine. At the time, the late 1990s, we were not alone in this quest as a prosperous economy enabled many wine lovers to seek out their little corner of paradise with a vineyard or winery of their own. The ensuing boom would have far-reaching consequences, as vineyard acreage in California expanded dramatically in the 1990s and early 2000s. Wine grape acreage in Sonoma County, for example, grew from thirty-three thousand acres in 1990 to fifty-eight thousand in 2006, while Napa County's vineyard acreage increased from thirty-two thousand acres in 1990 to forty-five thousand in 2006.

Terrific wine is grown in many places around California, most of which are stunningly beautiful places to live and work. Our personal

quest led us first to the well-established wine regions of Napa and Sonoma, both of which were within a couple hours' drive from our home
in Silicon Valley, just south of San Francisco. In the end, we were won
over by the beauty and agricultural diversity of Sonoma, its vibrant
food and wine culture, its proximity to San Francisco, and, of course,
its rich history of fine-wine production dating back to the middle of
the nineteenth century.

Like many novice winegrowers, we set out initially to find an established vineyard. It seemed like a faster, easier way to enter the business. Buying an operating vineyard would enable us to get a vineyard
with a known track record, produce (nearly) instant cash flow, and
save us the time, complexity, and investment of developing a vineyard
from scratch. Hopefully it would also come with a vineyard management team in place that could manage the property for us while we
continued to work in our existing careers for a few more years. Maybe
even an existing contract with a winery to buy the grapes. Sounds simple enough.

So we went in search of a small vineyard, something in the neighborhood of five to ten acres. Initially, it seemed that anything much
larger would be a daunting responsibility, if not prohibitively expensive, whereas smaller vineyards did not seem to be commercially viable. We wanted something that would be more than just a hobby vineyard but not too large. At the time, we were living about a hundred
miles south of Sonoma, working full-time, and eight to ten years away
from the prospect of quitting our day jobs and living full-time in the
Wine Country. A turnkey operation just seemed like the right choice.

We soon discovered several limitations and shortcomings in our
plan. In 1997, which was early in the 1990s vineyard boom, five- to
ten-acre vineyards for sale were few and far between. Zoning and land-
use restrictions had sharply curtailed the subdivision of large acreage
parcels into smaller properties of the scale we were looking for. Also,
most vineyards we found for sale were larger than ten acres, which was
thought to be the low end of what was viable to make a living growing

wine. Perhaps most significantly, it seemed that nearly all of the vineyards we found for sale were fraught with peril. Some were just out of our price range, often including very nice homes that drove the property value up. In other cases, disease or old age had diminished yields down to the point where the economics were not attractive, and the additional expense of replanting seemed inevitable. There were some properties, mostly old-vine Zinfandel vineyards, where ripping out the old low-yielding vines just seemed wrong, if not immoral. In any case, the economics of these old vineyards, with yields of three-quarters to one ton of grapes per acre, just never seemed to pencil out.

After a while, we reset our sights on virgin vineyard land, ultimately acquiring twenty acres of mountainous woodlands and meadows in late 1999. As the property was landlocked and completely undeveloped, we had to cut in a road, bring in electricity, and drill a well nearly four hundred feet down for water before we could put any vines in the ground. That work and the complexities of navigating new regulations on hillside vineyards added a couple of years to our timeline.

Whereas initially we had hoped to carve about five acres of vineyard out of the property, the limitations of our steep, rocky terrain led us to dial back our expectations just a bit further. So it was that in 2002 we converted four acres of south-facing meadows to vines. Even at that, the rather severe terrain forced us to plant these four acres in two blocks, separated by a steep, wooded ravine. The upper vineyard, as we call it, provides a beautiful scenic hillside just above where our house is now situated, a panorama we enjoy daily from almost every window on the front of our house. It was initially planted to about two and half acres of Syrah and a few rows of Cabernet Sauvignon, though some of that Syrah was later grafted over to Grenache. The lower vineyard, which is downhill and through the woods from our house, was planted to about an acre of Syrah.

In hindsight, scaling back our ambitions from ten acres to just under four proved to be a blessing in disguise. We now know that ten acres is way more vineyard than we could possibly manage ourselves.

A view from the upper vineyard looking down toward the house. (© 2012 by John Kiger.)

We were in search of a lifestyle, one that did not include managing employees but did embrace an active life managing the vineyard ourselves. Also, as we were soon to discover, the wine market is highly cyclical, and the financial responsibility that comes with a larger vineyard is much greater than we really want. After all, despite all the hard work involved, we really do like to claim we're "retired."

At the time we set off on this adventure, I had no idea that of the many educational and intellectual challenges I would face as a winegrower, I would find powdery mildew and all that surrounds it to be the most fascinating. This is quite an achievement for something so obscure and mundane, particularly as there seems to be no end to the new challenges, life lessons, and amusing anecdotes that life on a vineyard produces. On top of everything Mother Nature could find to throw at us as novice vineyard owners, we added an ever-changing cast of wooly, furry, and feathered creatures to an already overwhelming experience. We brought in sheep to "mow" the vineyard in lieu of herbicides, a livestock guardian dog to protect the sheep from mountain lions and coyotes, vineyard cats to keep down the voles that girdle and kill the vines, and free-range chickens for eggs and entertainment. Each new season in the vineyard and every addition to the menagerie came with interesting new challenges and rewarding experiences, but year after year one challenge would rise anew. A shape shifter, it changes from year to year. It lurks in the shadows in hot, dry years, only to explode and race through the vineyard like wildfire in a dry forest during cool, damp years.

What follows is the story of a winegrower and an organic philosophy that guides the annual struggle to coax great wine from a steep hillside and a few thousand vines. The story highlights, in particular, the many hazards of nature that lie hidden in any vintage, along with the trials and tribulations they bring. First and foremost among these hazards, grape powdery mildew is the star in a cast of scoundrels that threaten to wreak havoc in the vineyard at any time. What are these hazards, where do they come from, and how do they impact the people and vineyards that bring us the wines we all love? It is the story of the perennial fight to save wine from the ravages of nature.

The story starts in the next chapter, "The Origin of Our Affection and Its Peculiar Afflictions," where I set the stage with a trip back in time, describing the origin and history of wine grapes and the wine and cultivated vineyards that followed. I show how evolution and geography conspired to keep wine grapes apart from powdery mildew and other major wine pests for millions of years, and how once their paths crossed, neither would be the same again.

"It's Personal—Man versus Fungus" picks up the story in modern times, set against the backdrop of our early years as full-time winegrowers. Many lessons for the novice winegrower are revealed as we learn the ins and outs of managing our own vineyard and sow the seeds for our transition to organic farming. Here I also unveil the origins of my fascination with powdery mildew.

Chapter 4, "Revenge of the Fungi," introduces vineyard enemy number one, along with several other fungi that regularly wreak havoc in the vineyards of California and around the world. Questions answered include: Just what is powdery mildew? How do epidemics of powdery mildew occur in wine grape vineyards? How big a problem can a few fungi be? After all, it's kind of like a mushroom, right?

As if these pesky fungi were not bad enough, the fifth chapter spins a broader tale of death and destruction that awaits the unsuspecting vineyard owner. Chapter 5 describes many of the bacteria, viruses, and assorted animals that wreak havoc in our vineyards and describes both the agony and the ecstasy we have experienced battling these demons in our vineyard. As the old saying goes, if it's not one thing, it's another.

In the face of all these hazards, one might wonder why anyone would dare to grow wine grapes. Chapter 6, "Some Things Get Better with Age," shows how nature can rise to meets its own challenges, in this case focusing on the grapevine and how it responds to attack by powdery mildew and other fungal invaders. Despite an evolutionary path mostly devoid of exposure to powdery mildew, the grapevine nonetheless develops resistance to powdery mildew anew each year, albeit well after a period of naked susceptibility.

In the next chapter, "The Winegrower's Challenge," we examine how different farming philosophies play into the winegrowing challenge. What is conventional farming? What about sustainable farming? Organic and biodynamic farming? This chapter explains each of these farming philosophies in broad terms, as well as some of their more specific implications for important winegrowing practices like fertilization and pest control.

Chapter 8, "Going Organic, or Something Like That," chronicles our own transition from absentee farming to hands-on organic farming. This is a story of mountain lions, sheep, lambs, dogs, cats, squirrels, and mice, along with lots of compost as we jettison conventional fertilizers and pesticides in search of an organic ideal. And live to tell about it, somehow.

Powdery mildew and other vineyard pests do not take our attempts to control them lying down. In "The Art and Science of Mildew Control," we take a short tour though the basics of mildew prevention and control in modern vineyards. I explain the different types of materials that can be used to thwart mildew and other pests in conventional, sustainable, organic, and biodynamic vineyards and how growers decide when and how often to use different types of pesticides. And yet, despite our best efforts, mildew and other ravages of nature persist. In "Nature Strikes Back" we look at how powdery mildew, as a case in point, responds to our persistent efforts to eliminate it from our vineyards. As we have found with antibiotics in human medicine, excessive use or misuse of pesticides can lead to the development of pesticide-resistant strains of the pests we are trying to control. I show how winegrowers deal with pesticide resistance and the advantages of organic farming for avoiding the development of resistant strains of mildew and other pests.

"It's All-Out War" takes us back to our own ranch, Kiger Family Vineyard. In this chapter we look back over one of the worst powdery mildew seasons that California has seen in decades. This is a firsthand account of our monumental, season-long battle against a raging mildew epidemic, an experience that will imprint the memories of 2010 in our minds for years to come.

By this time, we will have seen how powdery mildew and other pests take a toll on both the vineyard and the farmer, so we turn to the reason we're doing all of this in the first place—the wines. In chapter 12, "From Vine to Glass," I show how powdery mildew, botrytis, and other fungal diseases, along with fungicide usage in the vineyard, affect winemaking and the wines we drink. The questions posed include: How does the battle to control these pests affect the wines we drink? Are there perceptible effects to consumers and wine enthusiasts? Are there other implications of fungicide programs we should be concerned about?

Moving closer to home, "The Odyssey in a Bottle" chronicles our own story of vine to glass. I describe how our philosophy of winegrowing is built on the idea that making our own wine closes a loop that starts in the vineyard and how the wine we produce came to be called The Odyssey.

The penultimate chapter, "Looking Forward," peers into the crystal ball, seeking to uncover clues to the future of pest control in wine grape vineyards. We look into ongoing research efforts to reduce fungicide usage by developing more resistant vines through genetic engineering and hybridization. In parallel, research continues into safer, more effective fungicide alternatives for both conventional and organic growers.

Finally, in "Truce," I close with some personal observations and a snapshot of a winegrowing lifestyle driven by the seasons and the vicissitudes of nature. We revel in the anticipation bud break brings each spring, the sigh of relief set forth by veraison, and the excitement of harvest that brings each vintage to a close.

ב

THE ORIGIN OF OUR AFFECTION
AND ITS PECULIAR AFFLICTIONS

WHERE DID WINE COME FROM TO BEGIN WITH?

W ine is at once both a simple prod-
uct of nature and a mystical product of human ingenuity. For millennia, wine has provided basic sustenance in many cultures, for the mind
if not the body, and has often been consumed daily as a dietary staple.
From these humble origins wine has evolved into a subject of considerable fascination, often romanticized and even the object of desire for
wealthy collectors. That wine can be all of these things certainly adds
to its allure. This fascination draws more than 7 million visitors a year
to the vineyards and wineries of Sonoma County in California and se-
duces wine collectors into paying tens of thousands of dollars for rare
bottles of wine at auction.

The origins of wine are shrouded in mystery, as befits such an ancient and revered icon of human civilization. Wine almost certainly
predates the cultivation of grapes and may well have provided the mo-
tivation for the domestication and cultivation of wild grapes in the first

place. While wine can be made from almost any type of grape, as well as other fruits such as apples, peaches, and blackberries, one particular species of grapes has dominated wine production throughout recorded history. This species is *Vitis vinifera*.

There are more than sixty species of grapevines in the Vitis genus, and most are native to the Northern Hemisphere in Europe, Asia, and North America. The native territory of *Vitis vinifera* lies between the northern latitudes of thirty degrees and fifty degrees in eastern Europe and western Asia. *Vitis vinifera sylvestris*, the wild predecessor to the wine grapes we know today, grows east to west from Lebanon to Spain, along the Danube and Rhine Rivers in its northernmost territory, and south to the headwaters of the Tigris and Euphrates Rivers.

Today nearly all of the world's major wine-producing regions lie between thirty and fifty degrees latitude in the Northern and Southern hemispheres, as these latitudes seem to define the natural boundaries for climates that are conducive to growing grapes for wine.

Believed to be somewhere between 130 and 200 million years old, *Vitis vinifera sylvestris* provided a tasty and nutritious food to the early human foragers who migrated from Africa into the Arabian Peninsula and eastern Mediterranean, possibly as far back as 100,000 to 125,000 years ago. Although no one knows for sure, the discovery of wine almost certainly came much later. Like many of the great human discoveries, the fermentation of grapes into wine was probably discovered quite by accident.

Fermentation is a naturally occurring process that gives us many of our most ancient and revered foods, including pickles, sauerkraut, yogurt, bread, beer, and wine. It is likely that all of these were discovered when foods were stored away or left lying around for fermentative yeasts or bacteria to act on, producing a new and transformative food or drink.

In the case of wine, it is not hard to imagine how grapes gathered for consumption as food were horded or left unattended, only to be found weeks later transformed into a soupy, mildly intoxicating brew.

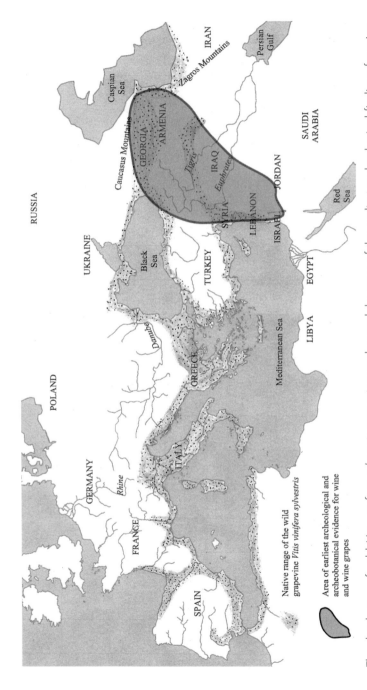

The distribution of wild *Vitis vinifera sylvestris* grapevines today and the area of the earliest archeological findings for wine-making and wine grapes. (Adapted from "Blank Roman Empire.png" by User:ColdEel at commons.wikimedia.org. Licensed under CC BY-SA 3.0.)

Archeological studies suggest that this discovery likely occurred some-where around nine to twelve thousand years ago. While humans were most certainly consuming grapes long before that, the earliest known man-made vessels or pieces of pottery in which *Vitis vinifera sylvestris* grapes could have been stored and (accidentally or intentionally) fer-mented date to about 7000 BC. Of course, it is possible that naturally shaped stone vessels or sacks made from animal skins were used to store food and drink prior to the invention of pottery, which could push back the discovery of wine thousands of years. We may never know for sure.

While there is no direct evidence of winemaking or storage in stone vessels, archeological findings provide the earliest direct evidence of wine in the form of residues on pottery shards found in the Za-gros Mountains of Iran, dated to around 5400 BC. This corresponds roughly to the late Neolithic or Stone Age period, when agriculture for food production was widespread in the Fertile Crescent. The core of the Fertile Crescent includes the Tigris and Euphrates River valleys, lying roughly between the eastern Mediterranean and the Persian Gulf, with the headwaters of both rivers extending northward into the Za-gros Mountains.

Early Neolithic agriculture was limited to grain crops, with fruits and nuts gathered from native, wild-growing plants. But it was in this period that the critical prerequisites to winemaking and growing occurred, in-cluding communal living in villages, the invention of pottery, and the discovery of various food-production and -preservation technologies.

It appears that grapevine cultivation originated sometime around 4000 to 6000 BC in the area between the Black and Caspian Seas, below the Caucasus Mountains. Today this area lies in the countries of Georgia and Armenia. Grape pips (i.e., seeds) identified by DNA analysis as *Vitis vinifera sativa* have been isolated in archeological sites in Georgia and Armenia dating back to around 5000 BC.

Vitis vinifera sativa is the cultivated descendent of *Vitis vinifera syl-vestris*, with *sativa* being Latin for cultivated. Archeological evidence and modern DNA analyses suggest that nearly all of today's wine grape cultivars are descended from *Vitis vinifera sylvestris* as variants of *Vitis vinifera sativa*. More than seven thousand cultivars, or cultivated wine grape varieties, have been cataloged in the *Vitis vinifera* family, which includes all of the familiar varieties, such as Cabernet Sauvignon, Chardonnay, Pinot Noir, and so on.

This early prehistoric cultivation of *Vitis vinifera* resulted in a few significant differences between the wild and cultivated variants. Most significant was the shift from sexual reproduction in wild grapevines to asexual reproduction in the cultivated varieties. In the wild, *Vitis vinifera* exists as separate male and female plants and requires both plants to pollinate grapevine flowers and produce fruit. In contrast, the cultivated variant is asexual, or hermaphroditic, where each flower contains both the male and female parts and is self-pollinating. The genetic mutation that allowed asexual reproduction was identified implicitly in the self-pollinating vines and propagated by early winegrowers.

A second interesting characteristic of the wild grapevine is that its grapes are normally dark skinned, or what we would call red grapes. Genomic analyses of red and white grape cultivars have determined that white grapes are the result of mutations in adjacent genes in a single chromosome, which essentially turn off the production of the anthocyanins that give red grapes their distinctive color. White grapes are rare in wild *Vitis vinifera sylvestris*, and only through vegetative propagation, most likely by rooting grapevine cuttings, were prehistoric growers able to maintain and propagate different types of wine grapes.

These early viticultural endeavors predate any recorded history and are thus inferred from various archeological disciplines, including from traditional archeology, paleontology, archeobotany, and biomolecular archeology. The first recorded histories or depictions of winemaking and winegrowing have been traced back to around 3000 BC in the eastern Mediterranean.

The Egyptians left extensive pictorial records of viticulture and wine consumption, many of which date from as early as 2400 to 2700 BC. Early Egyptian depictions provide our first insights into early viticulture, showing trellised vineyards growing around houses, among other plants, as well as stand-alone vineyards. Egyptian depictions of winemaking also include picking grapes into wicker baskets and stomping grapes by foot. Presses were made from cloth bags, which were attached to ropes stretched between poles and twisted to extract the juice.

As most of Egypt lies outside the ideal winegrowing regions between the northern latitudes of the thirtieth and fiftieth parallels, it is likely that Palestine, Syria, and Lebanon produced much of the wine consumed in ancient Egypt and Mesopotamia during the second and third millennia BC, being climatically and environmentally more suited to viticulture than the hotter, drier regions of Egypt and Mesopotamia.

From this eastern Mediterranean cradle of the wine grape, viticulture spread westward, possibly reaching Greece and Crete as early as the fifth millennium BC. Greeks appear to have been influential in the spread of viticulture throughout Mediterranean Europe. Theophrastus (c. 370–285 BC) provides one of the first detailed accounts of viticulture in Greece in his *Enquiry into Plants*. He describes vine physiology, methods of pruning and propagation, soils, pests, diseases, and environmental factors in grape growing.

Indications are that viticulture had reached Italy by around 900 BC, likely in Calabria and Campania. By 600 BC the seafaring Phocaean Greeks had spread viticulture to southern France in the area around Marseille. From here viticulture was spread north and west over the ensuing centuries, largely by the Romans, becoming well established in Bordeaux and the Rhône valley by the first century AD, and then spreading further north into France and Germany over the next few centuries.

It was the "Age of Exploration" and European colonialism that spread viticulture and wine to all corners of the globe. The Spanish brought viticulture to Latin America in the sixteenth century and ultimately into California in the eighteenth century. The Dutch carried it to South Africa in the seventeenth century, and the English to Australia and New Zealand in late eighteenth and early nineteenth centuries. English attempts to establish *Vitis vinifera* in the eastern United States in the seventeenth century were largely unsuccessful, owing to the rather harsh climate (hot humid summers, freezing winters), as well as to a variety of pests and diseases to which the vines of Europe were particularly susceptible, as we shall see in great detail shortly.

By the middle of the nineteenth century, cultivated wine grape vineyards had spread to nearly all of the climatically compatible temperate lands of Europe, North and South America, Africa, and Australia and New Zealand. Nearly all of these vines were *Vitis vinifera*, directly descended from the earliest cultivations from wild grapevines in the Near East thousands of years earlier.

NO GOOD DEED GOES UNPUNISHED

The implications of this genetic lineage and the methods used to propagate and multiply these vines had not caused the slightest ripple in the worldwide spread of viticulture into the middle of the nineteenth century. Nearly seven thousand years of viticulture, and all was well. However, the forces of colonialism and Darwinian evolution would conspire to wreak havoc on the world of viticulture in the nineteenth century and lay the foundation for significant changes to basic viticultural practices that endure to this day.

In hindsight, the first hint of trouble to come was the great potato blight that struck Ireland in 1844. By 1846 the blight had wiped out nearly the entire potato crop in Ireland, as well as in much of Europe, and reduced the population of the island nation by nearly one-fourth due to starvation and emigration. The culprit was a fungus, *Photophthora infestans*, native to South America and likely imported to Ireland accidentally on a transatlantic potato shipment. A paucity of genetic diversity in the Irish potato crop combined with susceptibility of the lumper potato variety favored by the Irish and a cool, damp climate to create the perfect conditions for the disaster.

With increased specialization of agriculture in the burgeoning industrial age and the influence of a global maritime trade, the spread of exotic pests and diseases was to become an acute problem for agriculture in the nineteenth century. The vineyards of Europe were no exception. In 1845 an English gardener named Edward Tucker discovered a dusty, white fungal growth on the grapevines in the garden he managed. The affected leaves would curl up and die, and the infection spread to the nearby grape clusters, which would themselves be covered by the same powdery white fungus before withering, cracking open, and decaying on the vine.

At the time of his discovery, Edward Tucker was familiar with the work of a certain Reverend M. J. Berkeley, who had previously identified the fungus responsible for the potato blight. Tucker sent Reverend Berkeley samples of his diseased vines for investigation. Reverend Berkeley christened the fungus he identified on the grape leaf samples *Oidium tuckeri* in Edward Tucker's honor but could offer no insight into the origins or control of the fungus.

Oidium tuckeri was almost certainly transported from America in the 1830s and 1840s on ornamental vines, which were popular in European gardens at the time. Native to North America, the fungus had coexisted for millennia with the native grapevines of the continent. As evolution would have it, most native American species of grapevines were naturally resistant. Though a grayish, dusty fungal growth had reportedly been found on cultivated *Vitis vinifera* vines in the 1830s in the United States, not much had been made of it, and no one foresaw its jump across the Atlantic; nor was it cause enough for concern that anyone bothered to look into its control.

In 1846, a year after Tucker's discovery of the dusty white fungus in his garden, the oidium appeared in France on grapevines in the kitchen garden at the Palace of Versailles. By 1851 the fungus had spread throughout France and was causing severe damage in the Languedoc wine region of southern France. It had also spread to the vineyards of Algeria, Greece, Hungary, Italy, Spain, Switzerland, and Turkey. The impact on wine production in Europe was staggering. In France alone, wine production declined from nearly 1.2 billion gallons in the early 1840s to less than 300 million gallons by 1854.

There was no precedent for such a destructive disease of grapevines anywhere in the wine-producing world at that time. The reaction in Europe was predictably hysterical. All manner of causes and treatments were offered. Causal explanations ranged from the wrath of God to poisoning of the soil by the laying of railroad tracks and telegraph lines across the continent.

Fortunately, there were others who saw similarities between the oidium of the vine and mildew infestations that afflicted other plants, most notably fruit trees. For example, a lime-sulfur mixture had been used to treat fungal infections on fruit trees as early as 1802. In 1846 an English gardener by the name of Kyle reported successfully treating oidium-afflicted vines with a solution of pulverized sulfur in water. At about the same time, Edward Tucker, who first reported the fungus, found that a solution of lime and sulfur could be used to control the oidium infection on his vines.

But successes were isolated, and reports of them were not widely accepted, particularly among the poor and uneducated peasantry that made up much of the winegrowing population of Europe at the time. Even among those who recognized the value of sulfur for controlling the oidium, the challenge of how to apply the sulfur on a scale of tens or hundreds of acres of vines was daunting. In 1850, a French gardener by the name of Gontier devised a bellows that enabled him to spread pulverized sulfur dust onto the moistened branches, leaves, and fruit of his vines. Despite his success, Gontier's method involved wetting the grape leaves and fruit and proved impractical on a large scale. In some cases, the morning dew could be used to good effect, but overall the solution remained impractical.

It was in 1853 that a French winegrower by the name of Charmeux devised a solution to the oidium epidemic. Charmeux found that sulfur dust itself, without wetting the leaves or fruit, was a sufficient control to prevent the onset or growth of the oidium if applied at opportune times. This finding, along with the development of handheld bellows and other sulfuring machines, enabled vineyard workers to apply the sulfur on a commercial scale and finally brought the European oidium epidemic under control. Nonetheless, applying twenty pounds or more of sulfur per acre using handheld or knapsack-style sprayers was hard work and very time-consuming. The oidium epidemic would soon be contained but would never truly disappear.

A LITANY OF PLAGUES MARCHES ON

The late nineteenth century proved to be a very difficult period for the winegrowers of Europe. The cultivated vineyards of Europe, having existed for thousands of years in relatively harmony, were to experience one crisis after another for most of the latter half of the nineteenth century. Oidium may have been the first, but by no means was it the worst.

In late 1867 and into 1868, published reports began to surface of a strange new malady afflicting vineyards in various wine regions of France, most notably in the Rhône valley. The leaves on seemingly healthy vines would begin to turn yellow and red in May or June and by late summer would die and fall to the ground. The grapes failed to ripen, and ultimately the affected plants would wither and die. Typically this malady would initially affect only a few vines in a vineyard but subsequently spread throughout the vineyard in all directions and to neighboring vineyards as well.

Mysteriously there were no signs of insects or fungal infections on the leaves, fruit, or woody sections of the vines. Jules Emile Planchon, a professor of botany at the University of Montpellier, dug out several of the afflicted vines and found extensive damage and decay to the root systems of the vines. On further inspection with a magnifying lens, he found eggs and adult female aphids, which he referred to as plant lice or root lice and which came to be known by the botanical name *Phylloxera vastatrix*.[1] The phylloxera would feed on the vine roots, sucking sap from the roots, and ultimately the injuries to the vine roots would cause the roots to die and the vine to wither.

It was soon discovered that the phylloxera moved from vine to vine in the soil but also spawned winged females that emerged from underground and spread in all directions on the wind. Over the course of the next two decades, the phylloxera infestation would spread to all the vineyards of Europe. The impact on wine production and vineyard acreage was devastating. In the 1890s there were some 1.5 million winegrowers in France whose livelihood depended on grape and wine production. At the time, grapes in France accounted for almost 15 percent of the total agricultural production of the country, exceeded only by dairy and wheat products. The devastating impact of phylloxera on wine production in France peaked in the late 1880s. In 1889, French wine production had fallen to around 630 million gallons as compared to 1.8 billion gallons in 1869. This was due to the reduced production of diseased vines and, in many cases, the abandonment of entire vineyards.

The culprit, *Phylloxera vastatrix,* and eventually the cure both proved to have their origins in the Americas. The phylloxera turned out to be native to North America and somehow made their way to France, most likely on American grapevines imported into France for ornamental or gardening purposes. Native American grapevines had lived with phylloxera for eons, and many species had developed a natural resistance to the damage of phylloxera feeding on its roots.

---

[1]Through much of the nineteenth and twentieth centuries, phylloxera was primarily classified as either *Phylloxera vastatrix, Phylloxera vitifoliae,* or *Daktulosphaira vitifoliae.* Today most scientists use the designation *Daktulosphaira vitifoliae.*

Although some scientists recognized the potential in this resistance, it was many years before it was accepted as the solution to the phylloxera problem. As early as 1869 researchers had proposed grafting *Vitis vinifera* onto the roots of resistant American vines to create a resistant grapevine, one that would produce true *Vitis vinifera* fruit without susceptibility to the root-feeding pest. The proponents of this strategy were called the *Americanistes*.

However, the fear of contaminating *Vitis vinifera* wine grapes with the tainted flavors of American grapes was widespread. It was well known that none of the native American grapes were suitable for producing quality wine, in particular because of the peculiar foxy, musky flavors these grapes imparted to their wines. Growers and researchers resistant to the idea of using American rootstocks tried countless other remedies, including pesticides and flooding the vineyards to kill the insects. None of these solutions proved effective or sustainable. Eventually the *Americanistes*, led by Professor Planchon prevailed, and by 1900 the replanting of French vineyards with American rootstocks had enabled French wine production to return to pre-phylloxera levels. Eventually, phylloxera would spread to nearly all the winegrowing regions of the world, and today most of the world's wine is made from grapes growing on grafted rootstock from phylloxera-resistant North American vines.

As if all this had not been enough for the vineyards of Europe, the importation of American vines to develop the cure for phylloxera brought with it new problems to compound the misery of winegrowers struggling to overcome phylloxera.

In the late 1870s, widespread reports of a new grapevine affliction began to circulate. This new malady attacked all green parts of the grapevines, including the leaves, tender shoots, and fruit. Symptoms first appeared as a white, cottony growth in lesions on the leaves and shoots. Infected shoot tips and leaves would curl and eventually turn brown and die. The young berries were particularly susceptible, appearing grayish when infected, and would become covered with a downy felt of mildew, ultimately rotting on the vine.

In 1878, Professor Planchon and his colleague, Professor Pierre Marie Alexis Millardet from the University of Bordeaux, diagnosed this malady as a new fungal infection never before seen on the grapevines of Europe. The new fungal invader was called downy mildew, known botanically as *Plasmopara viticola*. Downy mildew, like phylloxera and oidium before it, was a native of North America. In the rush to find a solution to the phylloxera epidemic, thousands of phylloxera-resistant grapevines had been imported to Europe for the breeding programs to develop new rootstocks that could withstand the phylloxera invasion. Unbeknownst to anyone at the time, along with these imported American grapevines came the fungus *Plasmopara viticola*. Microscopic fungal spores in the form of invisible downy mildew infections were carried into France on the very grapevines that were imported to help solve the biggest crisis winegrowers had ever faced. As these imported grapevines were planted in the nurseries and vineyards of France, *Plasmopara viticola* spores were spread on the wind to neighboring vines and vineyards. And so a new crisis began.

By 1882 downy mildew had spread to nearly all the winegrowing regions of France and into the other major winegrowing countries of Europe. In some ways, the downy mildew epidemic was worse than that of phylloxera. Granted, downy mildew was not fatal to the infected vines, but wines made from infected grapes were often undrinkable. Wine made from mildewed grapes could unexpectedly turn sour in the bottle, which became a common problem during several bad mildew years in the 1880s.

On the tails of the downy mildew epidemic came another North American grape disease, black rot, caused by the fungus *Guignardia bidwelii*. Black rot was most likely introduced into Europe in the same manner as downy mildew—on American grapevines imported in response to the phylloxera epidemic. In 1885 Professor Pierre Viala of the University of Montpellier identified the new ailment on grapes sent to him from a vineyard in southern France. Common to grapes in the United States but heretofore unheard of in Europe, black rot would attack the leaves, shoots, and grape clusters. As the name implies, infected grape clusters would darken and shrivel into unripe, hard, black raisins.

While the phylloxera epidemic was laid to rest with the introduction of phylloxera-resistant rootstocks, no such strategy of resistance was available to combat these new fungal invaders laying waste to the vines and grapes. While some native American grapes were resistant to one or more of the fungal diseases, this resistance could not be transferred to *Vitis vinifera* through grafting. In fact, the success of grafting European wine grape varietals onto American rootstocks was predicated on the idea that nothing would change in the *Vitis vinifera* character or the resulting wines.

Unfortunately, the sulfur spray that was instrumental in arresting the oidium epidemic of the 1850s was no match for the downy mildew and black rot epidemics that came nearly two decades later. Professor Millardet, who first identified downy mildew in French vineyards, found the solution in a serendipitous discovery. Millardet noticed that vines along a pathway through the vineyard at Chateau Beaucaillon in the Medoc had a bluish-white coating, as though they had been sprayed with some chemical, and did not exhibit any of the mildew symptoms found in adjacent vines without the bluish coating. On further inquiry, the grower explained that this was a common practice in the area, spraying the vines with a copper acetate or a copper sulfate and lime solution to discourage pilferage from the vines along the pathway.

Professor Millardet soon determined that copper was the active ingredient killing the fungus and, through experimentation over the next two years, determined that a solution of copper sulfate and lime in water was the most effective deterrent for battling the downy mildew and subsequently black rot as well. This became known as the Bordeaux Mixture. What remained was the same problem that plagued attempts to scale the sulfur solution for treating oidium in the 1850s—how to apply the Bordeaux Mixture on a large scale to tens or hundreds of acres of vineyards. The bellows developed for applying sulfur dust were of no use for the aqueous Bordeaux Mixture. It would take another couple of years for the industry to develop workable solutions, which would come in the form of knapsack or backpack sprayers. Brass or copper tanks were devised that could hold two or three gallons of the mixture, with straps for carrying the tank on the applicator's back and a bellows or diaphragm pump for drawing out and spraying the solution onto the vines. Strenuous work, by any measure. In an 1888 report from the United States Department of Agriculture, researcher A. W. Pierson reported on his work with Dr. Edwin Bidwell of Vineland, New Jersey, who first identified black rot on grapes in the United States, and Professor Viala of France to develop a treatment for black rot using the Bordeaux Mixture. Pierson recommended treatment of five hundred vines with twenty-five gallons of the Bordeaux Mixture. With average planting densities in French vineyard generally exceeding one thousand vines per acre, the task of such a treatment with three-gallon sprayers was daunting—especially given the need to treat vines every three weeks or so during the summer growing season.

THE AGE OF FUNGICIDES

By 1859 the application of sulfur to combat oidium was firmly established across Europe, and the practice enabled wine production to return to pre-oidium levels. Similarly, by 1890, application of the Bordeaux Mixture had the downy mildew and black rot epidemics under control. Horse-drawn sprayers for sulfur dust and liquid preparations such as the Bordeaux Mixture dramatically reduced the labor burden by the late 1890s, but viticulture had been changed for good. The regular application of fungicides to control disease had become a permanent fixture of the viticulture landscape.

We now know that *Oidium tuckeri*, the fungal pathogen responsible for the oidium epidemic, is not an oidium fungus after all, and

3

IT'S PERSONAL—MAN
VERSUS FUNGUS

People come into farming in many ways, from many backgrounds. But no matter how romantic, mystical, or fascinating wine might be, winegrowing is still farming. There are vineyards here in Sonoma County that date back to the nineteenth century, and many families have been farming their lands since before Prohibition. Crusty old growers who learned from their fathers and grandfathers and have been using the same basic practices for decades farm some of these vineyards. Others are second-, third-, or even fourth-generation growers who grew up in the family vineyards but studied viticulture in modern, cutting-edge programs at leading universities such as the University of California at Davis (UC Davis) and Fresno State University.

My path to viticulture placed me squarely in the tabula rasa category. I had not grown up on a farm, and the vineyards that are now common in my home state of North Carolina had yet to supplant the tobacco fields of my youth. However, my partly rural but mostly suburban upbringing helped a little, from the experiences of a large backyard garden to a father who grew up farming tobacco and other row crops, especially when compared to Deb's thoroughly urban experience as a child and young adult.

I began my education in viticulture by taking extension courses at UC Davis, initially by myself and later with Deb at Santa Rosa Junior College. These courses provided a good foundation along with a host of useful reference materials, but they probably amounted to less than one semester in a matriculated viticulture program. Mostly I learned by watching, doing, asking questions, and reading voraciously (praise the Internet!).

Fortunately, the winegrowers and wineries of Sonoma are a fabulously open and sharing community, which was very helpful for newcomers like us. This could not have been more different from the Silicon Valley business community I had come from, with its security passkeys for entering buildings, employee nondisclosure agreements to limit who can talk about what, and even briefcase scanners in the most security-conscious (paranoid?) companies. I have found this openness among winegrowers and producers to be fostered by a recognition that we are all in this together. There are some sixty thousand acres of grapes in Sonoma, with about eighteen hundred growers and three hundred wineries, all of whom depend on and benefit from the worldwide recognition of Sonoma as a source of great wines. Growers and wineries come and go, but Sonoma as a place persists, and the value of its reputation as a wine producer depends on the shared commitment of all involved. I suspect this is true of all the great wineproducing communities.

Our vineyard was in its fourth growing season when we moved there in 2005 and took over all of the farming operations. Up to that point we had been living about one hundred miles south, ensconced in busy and demanding careers. Prior to the planting of our vineyard in 2002, I spent a great deal of my spare time reading about growing grapes, along with the practical and financial aspects of owning a vineyard, but still had no hands-on experience. That changed once our vines were in the ground, and we spent many weekends soaking up as much knowledge and experience as our limited time and skills would permit. However, the vineyard was much more than we could handle as weekend farmers, so for the first three years a local vineyard-management company ran all of the day-to-day vineyard operations for us.

These vineyard management companies are an essential cog in the winegrowing machinery. Vineyard owners large and small employ them to manage independent and winery-owned vineyards alike. In our case, they enabled us to be absentee owners in the run-up to our transition to full-time winegrowers, but in the larger picture they provide essential labor and expertise to vineyards of all shapes and sizes. We found them to be part of a fascinating local business ecosystem that revolves around growing and producing wine. In addition to managing thousands of acres of vineyards in the area, they do most of the new vineyard development and vineyard redevelopment as well. They also provide crews, for their clients and independent growers, to pick the grapes at harvest time.

Other specialists abound in this ecosystem as well. There are viticulturists, typically with advanced degrees in viticulture from schools like UC Davis or Fresno State, who provide technical expertise on various aspects of growing wine grapes. Many vineyard management companies and wineries employ viticulturists, and others provide independent viticulture consulting services to growers on retainer or as needed. There are pest-control advisors, trained specialists in pest identification and management providing their services to management companies, wineries, and independent vineyard owners. There are many types of specialty equipment providers as well. Some businesses specialize in the fences, posts, wires, and other hardware required to install and maintain a vineyard. Tractors, mowers, sprayers, and other specialized machines for winegrowers are sold by various other businesses in the area. Still others sell the fertilizers and pesticides used by organic and conventional growers. A similar, parallel ecosystem thrives on the wine-producing side of the equation. It took quite a while to figure out that all of these resources existed and how best to use them.

All in all, our preparation for full-time farming consisted of little more than five years of reading about growing grapes and three years of weekend jaunts to the vineyard to check on and study the work of the people we had hired to manage the vineyard for us. Realistically, that was probably not sufficient experience to safely strike out on our own, but that was our plan, and we stuck to it. We hedged our bets somewhat by retaining a viticulture consultant, but mostly we were on our own. We also had an ace in the hole in the person of Robert Biale, namesake and principal owner of the winery that would be buying our grapes, Robert Biale Vineyards. Robert, or Bob, as his family and friends call him, is a third-generation winegrower who has winegrowing deeply encoded in his DNA. When we first met Bob, he and his partners had built a successful winery that had grown to cult status as a producer of amazing Zinfandel wines, including Zinfandel produced from the family's home ranch that was planted in the 1930s. We were fortunate that we met Bob as they were expanding into Syrah wines, to be called Hill Climber Syrah, which fit perfectly with our very steep hillside vineyard. All of the grapes produced in our vineyard since our first harvest in 2004 have gone to Robert Biale Vineyards, and Bob has been an invaluable resource and partner in our efforts to scale the winegrowing learning curve.

Suffice it to say, we were constantly overwhelmed and behind the proverbial eight ball for all of that first full growing season in 2005. More than anything that came before or since, that first season kindled the flame of curiosity that drives me to question the why and how of all our viticulture practices. Nothing frustrated me more than hearing, "That's the way we always do it," in response to a question about a particular viticulture practice or task.

Winegrowers are probably no different from any other group of people engaged in a seven-thousand-year-old craft. Science and technology are always pushing the envelope, only to be tempered by conventional wisdom, practical experience, and superstition. If nothing

Label for the 2008 Robert Biale Vineyards Syrah from Kiger Family Vineyard fruit. (© 2010 by Robert Biale Vineyards.)

else, it makes for a diversity of opinions and lively discussions among all involved.

The vineyard calendar begins early in the year with pruning in the winter and ends in late fall with postharvest feeding and winterization.

The tasks on the calendar don't change much from year to year but may move in or out by two to three weeks, depending on the weather that season. I developed my own vineyard calendar after that first season, partly to insure that we never fell so far behind again. It has some fifteen categories of activities and twenty-five to thirty discrete tasks that may occur in any given season. Some of these tasks may take an hour or two, while others may take days or weeks. Some may be done once a year; others are repeated as conditions dictate over the course of the season.

Deb and I have developed a good rhythm and division of labor over the years for managing the vineyard. It's maybe a sixty/forty split, where I do all the work requiring the tractor or other heavy equipment and act as our viticulturist to set the agenda for what we do and when. We jointly do all the basic manual labor in the vineyard, from weed whacking to canopy management and pruning. Picking the grapes at harvest is the only task for which we bring in outside help. Within our shared labors, we have our respective strengths, as I tend to work much faster, but Deb is more thorough and exacting. Back in the office, I manage all the data pertaining to our vineyard operations, such as the vineyard calendar and a record of all our activities and materials used in the vineyard. Deb is the financial controller, paying the bills and tracking all of our expenditures. She also manages our website and Facebook page and sends out two or three e-mail updates each year to the several hundred friends, family members, and acquaintances who follow the escapades of our vineyard odyssey. It is truly a team effort.

When I said we took over all the farming operations upon moving to the vineyard in 2005, there did remain one task we continued to contract out. That was spraying the vines to protect against powdery mildew. The vineyard management company that had been managing our vineyard since its inception continued to do the spraying for us in 2005 and 2006, employing a pretty conventional mildew-control program. I avoided this particular task partly because I did not relish exposure to the various fungicide chemicals employed and partly because it was physically very hard work to spray the entire vineyard with a backpack sprayer. At that time we did not have a tractor or tractor-attached sprayer, and even the vineyard manager's crew used backpack sprayers because he was not comfortable with the idea of his employees driving a tractor on our steep hillside vineyard.

Every month during the growing season for several years I had been getting monthly bills that included all of the labor and materials for the mildew-control program. These included bills for seemingly exotic agrichemicals from the likes of BASF, Dow Agrosciences, and Monsanto. I had not thought much about this when we lived a hundred miles away, but once were living in and around the vineyard, I began to have second thoughts about spraying these chemicals in what basically amounted to my front yard. Although we continued on this program for all of 2005, the seeds of change had germinated, and by the next season we decided to begin a transition to organic farming.

Beginning in 2007 we were on a fully organic farming program, which eliminated the use of all synthetic chemical fungicides, fertilizers, and herbicides. I also decided that I would now call the shots on mildew control, including what fungicides we would use and how often we would spray, but still have the contractor's employees do the actual spraying. With the benefit of hindsight, I can now say that this move was rather brazen, given, for example, that I had never actually seen mildew on a vine at that point. But I had been studying the topic religiously for most of the past year.

Among other things, I had always been bothered by the routine of our mildew program, and by extension that of almost every other grower I knew. In conventional mildew-control programs that basically meant spraying every twenty-one days from bud break to veraison, mostly regardless of the weather. Veraison, which is the onset of ripening and is signaled by the color change in red wine grapes, typically begins about four months after bud break. With the transition to organic farming, that would translate into spraying every seven to fourteen days for nearly four months, as fungicides approved for organic production are less effective and less persistent on the vine.

What really bothered me was that in the course of studying the problem, I had learned that the weather during the annual growing season is not uniformly conducive to mildew. For instance, early in the season it is often too cold for mildew to grow, and these conditions can persist after bud break in some years for five or six weeks. Whereas the science said no spraying was necessary, common practice and almost every grower I knew sprayed anyway. Better safe than sorry was the motto. But that did not sit well with me, if for no other reason than that I did not like the idea of spraying fungicides unnecessarily, not to mention the added cost.

I decided that with science on my side, I could manage this better and more cost-effectively at the same time. Science in this case came in the form of a weather station that calculated a Powdery Mildew Risk Index based on the specific weather conditions in my vineyard. The Powdery Mildew Risk Index had been developed by plant pathologists at UC Davis and was based on the sensitivity of the mildew reproductive cycle to specific weather conditions. When the index is high, conditions for reproduction are favorable, and the interval between sprays needs to be shortened. When the index is low, conditions are unfavorable, and the interval can be lengthened (i.e., fewer sprays, further apart).

You have to remember that in 2007 I was still something of a novice grower. I regularly met and talked winegrowing with people who had forgotten more about growing grapes than the sum of all my knowledge. But on the topic of mildew control, something did not sit right with me. Maybe I was a bit naïve, but I felt then, as I do now, that we should do everything possible to minimize the chemicals we are spraying on our vines, for the health of the vineyard ecosystem and surrounding community in general and the consumers of our wines as well. Even when farming organically, we're still spraying fungicides. Granted, they may be softer in terms of impact on the environment and consumers, but they are not totally benign.

Let me be the first to say that, on the whole, winegrowers are conscientious and environmentally sensitive managers of the lands they farm. I have seen firsthand the changes in practices and the data showing how Sonoma County growers have reduced pesticide usage significantly over the past ten or fifteen years. So I am not suggesting wanton disregard for the environment, much less careless or excessive pesticide usage. Rather, I saw an opportunity to do better, to use science and data to improve on standard practices, and maybe to set an example for others in the process.

So this was how my obsession began. Could I do things differently and be successful? Maybe change the way others work in the process? I would need to be successful not only in my own vineyard but as an evangelist as well. The benefits would be concrete—less pesticide use, reduced tractor time and fuel usage, and lower labor costs. The costs would be less certain but daunting nonetheless—an increased risk of mildew infection.

Over the next three years, I developed and implemented a fairly rigorous program based on the Powdery Mildew Risk Index. I recorded the daily index readings from my weather station during the growing season and managed my spray program accordingly. I did not start spraying at bud break in the spring but waited until the index trigger indicated that the initial conditions for mildew reproduction had been reached. I also lengthened or shortened the interval between sprays and adjusted the quantity of spray materials used according to the disease pressure indicated by the index readings. During this period I also bought a tractor and sprayer and took over all the spraying operations myself.

Within the Sonoma grape-growing community, there are a lot of forums for growers to interact and exchange ideas. The Sonoma County Winegrowers, a grower-funded organization that provides advocacy, marketing, and educational support for Sonoma winegrowers, grapes, and wines, organizes monthly meetings during the heart of the growing season. One series is called Integrated Pest Management, which provides a forum for growers to discuss current pest-management issues in their vineyards. A second series is the Organic Producer Group, which convenes at a different vineyard each month, focusing on the organic farming practices, challenges, and successes of that operation. Powdery mildew control plays a big part in both forums.

I have been very active in both of these forums since my first year of full-time winegrowing. The well of knowledge and experience to draw on in these meetings is impressive. Participants run the gamut, from small farmers like myself to university-trained viticulturists from some of the largest vineyard management operations in the region. Both forums promote best practices in pest management, reduced pesticide usage, and a smaller, softer environmental footprint in general. In addition to soaking up all the knowledge I can from these forums, I also advocate for my approach to farming, not just for powdery mildew control but for low-input, organic farming in general. To that end, we hosted one of the Organic Producer Group meetings at our vineyard, which gave us the chance to talk about and get feedback on our farming practices.

Though I'm no longer really a novice, there is still much I haven't seen in the vineyard. And no matter how long I work at it, my small operation will never provide the breadth of experience that a viticulturist managing hundreds of acres sees each year. On the other hand, the small scale of our vineyard permits a hands-on approach that is not possible for owners of larger vineyards, and frankly it's more work than many owners of smaller vineyards are willing to shoulder. Where powdery mildew is concerned, that includes monitoring the Powdery Mildew Risk Index, selecting and buying the fungicides to use, setting the schedule of sprays, determining the spray rate and volume, setting up the tractor and sprayer to deliver the specified rate, measuring and mixing the materials, driving the tractor and operating the sprayer, cleaning the equipment, reporting pesticide usage to the agricultural commissioner, and monitoring the fields for mildew infestations.

After a few years of this, I became more confident that I was on the right track. I'd hesitate to say I was smug about it, but I do enjoy advocating for my ideas as well as helping others with their winegrowing programs or problems. Of course, just when you think you know it all, nature throws you a nasty curve. One might say, it was the revenge of the fungi.

4

REVENGE OF THE FUNGI

Almost anyone who has done any gardening or landscaping has encountered powdery mildew in one form or another, as it is one of the most common pests afflicting the plants we grow. If you have ever grown roses or squash, you have almost certainly seen powdery mildew, starting with powdery white spots on the leaves that eventually cause the leaves to yellow and die. While it may not kill the plant, if left unchecked its effects on foliage, flowers, and fruit can be devastating.

Powdery mildew is commonly found in vegetable gardens, on everything from artichokes to zucchini, as well as many grain and fruit crops, ornamental plants, and even common shade trees. Table 4.1 shows a partial list of the many plants that are affected by the broad spectrum of diseases commonly called powdery mildew.

Powdery mildew is a fungus (plural: fungi). The most common fungi we encounter in everyday life are mushrooms and the molds we find on foods and various household surfaces. Powdery mildew growing on a plant is basically a type of mold. When a plant has powdery mildew, that means a fungus is growing on or even inside the plant's tissues. Despite its apparent ubiquity, powdery mildew is actually a

Table 4.1. Powdery Mildew Varieties

Vegetables	Artichoke, bean, beet, broccoli, carrot, cucumber, eggplant, lettuce, melon, parsnip, pea, pepper, potato, pumpkin, radicchio, radish, squash, tomato, turnip
Grains	Wheat, barley, oat, rye, triticale
Fruits	Grape, peach, nectarine, strawberry, cherry, apple, plum, apricot, raspberry, blackberry, blueberry
Trees	Oak, poplar, dogwood
Ornamentals	Roses, chrysanthemum, crape myrtle, rhododendron, azalea

plant-specific disease, with a unique species of mildew for nearly every host plant. The powdery mildew on your roses is not the same species that infects your cucumbers, both of which are different from grape powdery mildew. However, they all share common life cycles and reproductive traits, and are sensitive to the same types of controls.

All of the different types of powdery mildew are classified as ascomycetes in the fungus world. Ascomycetes are the largest division, or phylum, in the fungus kingdom and include many pathogenic and beneficial species. Beneficial species include the penicillium fungi, from which penicillin antibiotics are derived, and the saccharomyces yeasts used in baking, beer brewing, and winemaking. All of the ascomycetes fungi are characterized by a common reproductive mechanism, called the ascospore. Mushrooms, mildews, and molds that we encounter in everyday life all reproduce by releasing spores, which are spread as they are scattered by wind and other air currents. In the case of powdery mildew, the ascospore is the tiny egg-shaped spore that earned grape powdery mildew its rather ominous nickname: oidium.

Although the many diverse plants affected by powdery mildew may host different species of mildew fungi, their common family ties make the different species of powdery mildew susceptible to the same environmental influences and disease or infection controls. For example, many of you have probably noticed that some vineyards are decorated with rose bushes at the ends of the vine rows. No doubt this is decorative, and indeed it is generally found where the aesthetic impact is greatest, such as along roadways and around the houses and winery structures of the vineyard property. But there is a functional aspect as well, though it may be a bit antiquated in modern viticulture. It is likely that the practice of planting roses in vineyards arose in France in the late nineteenth century in response to the devastating powdery mildew epidemic of the 1850s, the idea being that roses are somewhat more susceptible to mildew than grapevines, and the appearance of mildew on the roses in the vineyard served to prompt the grower to treat the vines immediately. Nineteenth-century winegrowers lacked the sophisticated climate-based models of mildew growth and reproduction that we have today. Their mildew-control programs generally called for a limited number of preventive sprays to coincide with specific phenological events, such as bud break, flowering, and fruit set. Hence, early-warning systems for mildew outbreaks, such as infected rose bushes, could be very valuable. And the mildew in the rose bushes could be controlled with the same treatments that were used to treat the vineyard.

Powdery mildew, being a fungus, often leads us to think about it the same way we might think about other common fungi that we are more familiar with, such as household molds and mushrooms. Some of that thinking helps, but some of it is misguided. The most significant misconception is the idea that powdery mildew, like household molds and mushrooms, needs dark, wet conditions to thrive. While there is some truth to this, when we look more closely at powdery mildew, we will find that the most virulent outbreaks can occur in the arid, rain-free growing seasons of California, Australia, and other winegrowing regions where water is a scarce commodity during the height of the summer. Consider, for instance, that the average annual rainfall in Sonoma County for June and July combined is less than one-quarter of an inch. Yet this is the height of powdery mildew season in the vineyards of Sonoma.

The conditions favoring powdery mildew exist wherever wine grapes are grown. These conditions vary across the different winegrowing regions of the world, with some regions being more susceptible than others. Even within a single region, some vineyards will be more disposed to infection than others, just as some years will be worse than others due to different factors, such as sun exposure or the relative prevalence of fog, rainfall, or favorable temperatures. It is also true that some varieties of grapes are more susceptible than others. Some, such as Chardonnay and Carignan, are notoriously prone to infection. In general, all of the *Vitis vinifera* varieties are susceptible and require some type of program to manage or prevent mildew infection.[1]

DELICACY OR DISEASE?

It's ironic that fungi are capable of providing such wonderful delicacies as truffles, sourdough bread, and wine, not to mention life-saving medicines like penicillin, while at the same time wreaking havoc on all manner of plants and animals through a host of infectious diseases. Athlete's foot and yeast infections are common fungal diseases that plague humans. While all too common, these are not particularly serious illnesses. In the grape world, on the other hand, fungi are responsible for a broad spectrum of very serious diseases. Powdery mildew, while the most common and widespread, is hardly alone.

As described earlier, downy mildew and black rot followed the powdery mildew invasion of European vineyards in the late nineteenth century, leaving considerable devastation in their wake. By some accounts, downy mildew is the most serious and destructive grapevine disease. Although not as widespread as powdery mildew, it can be more difficult to control, especially for organic growers. Whereas powdery mildew thrives even in the arid winegrowing regions of the world, downy mildew is only a problem where rainfall is a common occurrence during the growing season. While many important winegrowing regions in places such as California, Australia, and Argentina

---

[1]Details on the biology and symptoms of grape powdery mildew infections can be found in the appendix.

experience little or no rainfall during the summer, other important winegrowing centers rely on rainfall to water their vineyards in the summer. The vineyards of France eschew drip irrigation in favor of Mother Nature's rainfall, but at a steep price. These vineyards, as well as many others in Europe and most of the United States, are under near constant threat of downy mildew infection. To make matters worse, the same weather conditions that give rise to downy mildew also promote black rot. Black rot is an ascomycetes fungus like powdery mildew, but unlike powdery mildew, it is unable to thrive in arid conditions. Black rot is an enormous challenge for organic winegrowers in the eastern United States and in some parts of Europe as well, but it is practically unknown in California.

There is, oddly enough, one fungus that provides both delicacy and disaster in the wine world. Formally known as *Botrytis cinerea*, the botrytis fungus is sometimes referred to as the "noble rot" and is highly prized for its role in producing some of the great dessert wines of the world, including the Beerenauslese and Trockenbeerenauslese wines of Germany and the Sauternes of France. Under ideal conditions, the fuzzy gray mold of botrytis-infected fruit will desiccate the grape berries, concentrating fruit sugars, acids, and other flavor compounds to produce a wonderfully sweet and complex wine. But these conditions are rare. More commonly botrytis is the last straw in a challenging or difficult vintage, with significant crop losses as botrytis bunch rot consumes infected grape clusters. In California, botrytis infections generally occur following early-season rains, infecting grape flower blossoms and immature grape clusters. In other areas with more summer rainfall, infections may also occur during the summer, although the fungus has trouble penetrating the tough green berry skin without some type of injury, such as insect or bird damage. These infections lie invisible and dormant until harvest time nears, awaiting the perfect storm of late-season rainfall on the soft, penetrable skins of ripe grapes. Botrytis risk is especially high in bad powdery mildew years, as conditions favoring mildew also promote botrytis, and mildew damage to the grape berries provides a welcome environment for the botrytis fungus. But more than anything, rain during the harvest season invites botrytis disaster, especially in tightly clustered varieties such as Chardonnay, Zinfandel, and Grenache. Unseasonably large rainstorms swept across Napa and Sonoma in October 2011, when much of the grape crop was still hanging in the vineyards because of an exceptionally cool summer and fall. The result was predictably disastrous, and large swaths of Cabernet Sauvignon, Syrah, and other later-ripening varieties were left to rot in the fields as botrytis bunch rot exploded across the region. We may have lost 2 or 3 percent of our fruit to botrytis that year, but other vineyards were left unpicked to rot in the field as botrytis tore through the vineyards in late October.

All of these fungal diseases have a number of things in common. Most notably, they all depend on actively growing grapevine leaves, flowers, or fruit to grow and reproduce. Also, while they can easily wreck a vintage with moldy, rotten fruit, they will not permanently damage the vines. Unfortunately, there also exists a host of fungal pathogens that present a polar opposite threat. These fungi attack dormant vines, infecting the woody material of the vine trunks and cordons. And they are deadly.

Collectively known as grapevine trunk diseases or canker diseases, these diseases are fungal infections that enter grapevines through pruning wounds during winter and early-spring rains. Grapevines are pruned annually during the winter dormant season to manage the annual growth, vigor, and fruiting capacity of the vines. In California this coincides with the rainy season, but rain is common during winter and spring in nearly all of the major winegrowing regions of the world. In our vineyard, most of the vines are pruned to eight or ten spurs from which the next year's growth will emerge, and each of these spurs has two pruning cuts or wounds that are potential entry points for infection. Although each cut is small in surface area, an acre of vines may have fifteen or twenty thousand such cuts. While the potential for any given cut to become a infected is relatively low, the probability of one

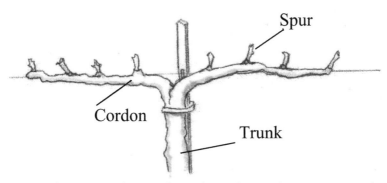

A pruned grapevine showing the trunk, cordon, and spurs. (Drawing by Sherry Rahn.)

or more vines being infected in any given winter is considerable, given the large number of these pruning wounds in every acre of vines. Unlike mildew infections, which can spiral out of control in a manner of weeks, these trunk diseases tend to spread slowly over the years but can build up to the point where vineyard productivity and lifespan are severely impacted.

The first sign of these diseases is often misshapen and stunted growth at one or more spur positions or dead spurs that do not "wake up" with the rest of the vine in the spring. Over time the stunted spurs will die, and more dead spurs will appear as the fungus spreads through the vine. If you take a cross-sectional cut of the vine near the dead or stunted spur, you will find a wedge of dead vine trunk that looks like a slice of pie. This dead pie slice in the trunk is the canker after which these diseases are named.

Each winter during pruning we find a few dead spurs. Sometimes you can cut the vine back far enough to eliminate the canker, in which case you have to grow a new arm (cordon) from a trunk sucker or a shoot on the other arm of the vine. In either case, you lose some production from that vine for a couple of years. If the canker gets down into the trunk, you can easily lose the whole vine. For many years we delayed our pruning until March, counting on the winter rains to spend most, if not all, of the reproductive spores before we started pruning in the spring. This has been shown to dramatically reduce infections from these trunk diseases, but it tends to delay bud break by a week or so, which can come back to bite you in the fall if it means you are harvesting a week or so late under threat of soaking rainstorms— which is exactly what happened in 2009, 2010, and 2011. So, after three years of late, rain-soaked harvests, we moved our pruning up to January, with the additional cost and labor of painting each pruning cut with a protective skin of latex paint. While there are some fungicides available to protect against these trunk diseases, none of these are suitable for organic production. Well, nobody ever said this was going to be easy!

5

IF IT'S NOT ONE THING,
IT'S ANOTHER

Here in Sonoma, vineyards along the low-lying, fog-prone stretches of the Russian River Valley battle powdery mildew all season long in most years, while just a few miles away, many of the warm sunny vineyards in Alexander Valley enjoy relatively low mildew pressure in all but unusually cool summers. Sometimes the combination of the local microclimate and the grape variety planted in a particular vineyard can combine to make powdery mildew an even greater threat. For example, many of the best Chardonnay wines made in California are grown in the fog-prone stretches of the Russian River Valley and Sonoma Coast regions, exacerbating Chardonnay's inherent susceptibility to powdery mildew. These same conditions, along with Chardonnay's tendency to produce tightly bunched clusters of berries, also make botrytis a perennial threat along the Russian River.

Our own vineyard lies somewhere between those extremes, with lots of morning fog that gives way to mostly sunny days, leading to only moderate mildew pressure in most seasons. In our case, the Syrah is least prone to infection, whereas the Grenache is a bit too welcoming as a host, and the Cabernet Sauvignon fits in somewhere between those two. I discovered these particular sensitivities the hard way, as it seems nature is inclined to educate farmers mostly through trial by fire.

Initially planted entirely to Syrah, save for two rows of Cabernet Sauvignon, I found nary a spot of mildew in the first seven years. I must say I was somewhat excited, if a bit terrified, when I found the first mildew growing in our vineyard in late July of 2009 (excited if for no other reason than finally verifying that I would know it if I saw it). Fortunately it was confined to a couple of Cabernet vines and easily snuffed out before it could spread to any other vines or cause significant fruit damage

Earlier that year we had grafted a block of Syrah vines over to Grenache. In the first growing season, the Grenache vines were relatively small, with good light and air penetration throughout the canopy, and bore no fruit. So whatever mildew terrors the Grenache held in store were kept well hidden for the time being. While particular vineyards and grape varietals may vary in their susceptibilities to mildew infection, nothing trumps a bad mildew season brought on by a long, cool, and foggy summer, which is exactly what we got in both 2010 and 2011. Our first two seasons of full-size, fruit-bearing Grenache vines coincided with two of the worst mildew seasons that Sonoma had seen in a generation. And so I learned firsthand the extent to which some grape varieties are more susceptible to powdery mildew than others!

In 2010 and 2011 winegrowers up and down California battled epic mildew infections. In both those years, long cool summers pushed harvests back several weeks, into late October and early November for many vineyards, which set up massive botrytis outbreaks after unusually strong October storms drenched the vineyards. Organic growers really struggled, armed as we were with less powerful tools than the synthetic chemical fungicides available to conventional growers. Of course, these trials and tribulations never occur in isolation. Leading up to 2010, we had experienced three years of drought in California. The drought years of 2007, 2008, and 2009 had resulted in relatively small canopies in the vineyards, due to low soil moisture as the vines launched into their most vigorous periods of growth each spring. Smaller canopies, together with warm, dry springs and summers, during those years reduced overall mildew pressure. Hence we were caught a bit off guard by the explosive canopy growth from late spring rains in 2010 and 2011, which combined with the generally cool and damp weather to produce those epic mildew seasons.

Even as the mildew battle ebbs and flows, a host of other pests and diseases abound. Insects, birds, bacteria, viruses, and fungal diseases threaten to rain calamity down upon the winegrower at any time. Not every grower faces the same threats, but every grower is susceptible to one or more. Some of these threats have a worldwide presence. Others are limited to specific geographies or climates.

Take phylloxera, for example, which nearly wiped out European grapevines in the nineteenth century. The phylloxera epidemic that started in 1860s France worked its way around the world in the ensuing decades. By the early twentieth century, most *Vitis vinifera* in the world had been replanted onto phylloxera-resistant rootstocks. Or so we thought. From the 1890s through the 1970s, most vineyard plantings in California used a rootstock created during the nineteenth-century phylloxera crisis. Called AXR1, it was a cross between a phylloxera-resistant *Vitis rupestris* grapevine from North America and a European *Vitis vinifera* vine. AXR1 performed admirably in California for decades, but in the 1980s a devastating wave of phylloxera infestation swept through the vineyards of California, including those planted on AXR1. By the mid-1990s, somewhere between one-half and two-thirds of the vineyards in Sonoma and Napa had to be replanted, at great financial expense and loss of grape production for three to five years. Today, phylloxera remains an ever-present threat, lurking just below the surface. It is not a pest that can be treated or managed once it moves in, so the best insurance against infestation remains the use of phylloxera-resistant rootstocks when planting a vineyard.

There are some pests and diseases that growers can manage with the proper viticulture practices and, where necessary, the judicious use of treatments such as fungicides or insecticides. Other pests are best avoided, such as phylloxera, as there are no treatments or management practices that will save your vineyard once infected. Among the most common and devastating of the untreatable afflictions are viral infections. There are a great many viruses known to infect grapevines. Many of the known viruses can be avoided by the use of certified virus-free vines at the time of planting. But others lurk unknown or undetected in the grapevine planting stock or even in neighboring vineyards awaiting an insect carrier to bring them across the fence into your vines. Some are deadly; others are merely debilitating. The debilitating viral infections tend to suppress vine vigor and delay fruit ripening, which may require replanting of the vineyard as wine quality declines or the vineyard becomes unsustainable as the yield declines below the cost of production.

The deadly ones, on the other hand, are just that—deadly. I know that because I have witnessed firsthand what can happen when such a virus strikes a vineyard. At the time we planted our vineyard, we chose certified nursery stock from three different grapevine nurseries. Certified nursery stock includes grapevines that are certified to be free of known viruses, grafted onto phylloxera-resistant rootstock by the nursery, and delivered to be planted as dormant, bare-rooted vines. These are much like the dormant, bare-rooted fruit trees you might buy from your local garden store to plant in your home garden.

We planted three different clones of Syrah: Syrah Noir, ENTAV 383, and ENTAV 877. As is customary, each clone was planted in a separate block rather than intermingled throughout the vineyard. In the grapevine world, clones are variants of a single variety of grape that are selected and propagated to promote certain traits, such as berry size or cluster size. All clones of Syrah are Syrah and are easily distinguished from other varieties, such as Cabernet Sauvignon, or even closely related varieties such as Petite Sirah. But they do exhibit subtle differences. In our vineyard, for example, the Syrah Noir produces very loose, open clusters with small berries. These are very good properties, as loose open clusters are less prone to fungal infections, and small berries provide a high skin-to-pulp ratio that is prized for high-quality wines. The yields, on the other hand, tend to be small as a result. The ENTAV 877 clone, by contrast, produces larger, tighter clusters with medium-sized berries. The higher yields and tighter clusters require a bit more work in the vineyard but also produce a high-quality wine.

Planted in 2002, the vines followed a fairly typical course for most of the first three years, bearing only a tiny crop in the third year. It was in 2004, however, when our vines set their first crop, that the signs of trouble began to appear in the ENTAV 877 block. Showing a full, lush green canopy in midsummer, leaves on some of the vines began to turn bright red before veraison. The red and yellow foliage you see in Wine Country in late fall makes for beautiful scenery and, once the grapes are ripe, will have little or no impact on wine quality. But when the leaves prematurely turn red earlier in the season, photosynthesis is reduced or stops all together before the grapes can ripen fully.

The situation worsened the following season, with many more vines turning red prematurely. Also, some of the vines that had turned red began to wither and die. By 2006 we were experiencing significant vine losses in the ENTAV 877 block, losing about 3 to 5 percent of the vines each year while losing the fruit on many more vines, as the red leaves were unable to ripen the crop.

We tried everything we could think of to stem the tide. We increased the water to the affected vines. We fertilized them individually, by hand. We reduced the crop load early in the season. Nothing we did had any effect.

I called the University of California Cooperative Extension viticulture advisor for Sonoma County, who brought in researchers from UC Davis and the United States Department of Agriculture. They took samples for inspection and tested for known viruses. For better or for worse, I was not alone. In the late 1990s and early 2000s, these symptoms had been reported on Syrah vines in France, California, and other regions of the world where Syrah was being grown. By 2007, the problem had grown to such proportions that an international conference was held in Davis, California, where researchers from the United States, France, and South Africa met to present and discuss their research on the causes of this mysterious disease, variously called Syrah decline or Syrah disorder. I attended this conference, of course, but came away with no assurances other than the fact that the problem was serious and not likely to get better.

By 2009 some answers began to emerge with the identification of two previously unidentified viruses in affected vines. One was a new variant of a common grapevine virus, called the Rupestris stem-pitting-associated virus–Syrah variant (RSPaV-Syrah), while the second was aptly named Grapevine Syrah virus–1 (GSyV-1). Much less is known about the Grapevine Syrah virus, but the Rupestris stem-pitting family of viruses is all too common and fairly well understood. Fortunately it is not thought to be contagious from infected vines to other nearby vines. The Rupestris virus is transmitted when the vines are propagated at the nursery.

There is little consolation in this information, other than the indications that the virus is not likely to spread from my ENTAV 877 vines to the other two Syrah blocks in the vineyard. In fact, the first row of the Syrah Noir vines adjacent to a row of ENTAV 877 vines looks completely different and continues to produce wonderful fruit. The ENTAV 877 vines are another story. With fewer than three-quarters of the vines still alive, and many more on the decline, they will soon be ripped up and burned.

Unfortunately, not all such viral predators of grapevines are so easily contained. Other grapevine viruses are spread from vine to vine, and even to neighboring vineyards, by insects. Two of the most serious such threats today are grapevine leafroll virus and fanleaf virus. Both leafroll virus and fanleaf virus suppress vine vigor and fruit ripening but are generally not fatal. The impact on vineyard economics and wine quality varies from site to site, but the resulting poor yield and wine quality forces many infested vineyards to be replanted.

Leafroll virus has been known in California vineyards since early in the twentieth century, but in recent years it has begun to spread at alarming rates. Long thought to spread only by propagation, as with the Rupestris stem-pitting virus, it is now known that leafroll virus is also spread by mealybugs. Mealybugs are scale insects that feed on vine foliage and fruit and are spread within and between vineyards on wind, equipment, and even harvested grapes. This is potentially a very serious problem, as mealybugs are nearly impossible to eradicate. Biological controls, along with viticultural and other treatment practices, can effectively manage the direct impact of mealybug damage to fruit and foliage. However, even small populations that would not otherwise cause notable economic damage may present a significant long-term threat as leafroll virus carriers.

Fanleaf virus symptoms were first described in the vineyards of France in the mid-nineteenth century, and this grapevine disease remains one of the most serious worldwide. Although the virus has long been known to be a soil-borne disease, it was not until around 1960 that the cause and means of transmission for the disease were identified. The culprit spreading the disease from vine to vine is a tiny roundworm, known formally as a dagger nematode, that feeds on grapevine roots. For decades, replanting of fanleaf-virus-infected vineyard sites has entailed fumigation of vineyard soils to eliminate the nematode populations. However, fumigation is dangerous for the vineyard workers and injurious to all soil-borne organisms, including the many beneficial organisms vital to a healthy soil food web. In recent years, nematode-resistant rootstocks have been developed, providing an alternative to fumigation for vineyard development in affected areas.

Not all insect pests in the vineyard carry the double whammy of fruit or foliar damage plus disease transmission. Many of the insect pests in vineyards are pests precisely because of the damage they cause to the vines, generally by feeding on the fruit or foliage. One significant exception to this is the sharpshooter, a small leafhopper that feeds on grapevines. Actually, there are several species of sharpshooters, but the glassy-winged sharpshooter and the blue-green sharpshooter cause the most damage in California vineyards. The sharpshooters carry a strain of bacteria called *Xylella fastidiosa* that causes Pierce's disease in grapevines (along with citrus, almond, and stone fruit trees). The bacteria are transmitted to the vine during the insect's feeding. The bacteria live and multiply in the vine's xylem, eventually blocking the movement of water and nutrients, thus killing the vine. Sharpshooters, and thus Pierce's disease, are very common throughout the southern United States, from Southern California to Florida, and may be found in any US winegrowing region with relatively mild winters, including Northern California's Sonoma and Napa Valleys. The blue-green sharpshooter is the primary cause of Pierce's disease in the north coast vineyards of Napa and Sonoma and is responsible for significant Pierce's disease damage in the region since the mid-1990s.

Regular monitoring and treatment for insect pests such as mealybugs and sharpshooters is generally required only when signs of infestation are present. I have not experienced any insect-related problems in my vineyard, and so, until recently, I did not maintain any regular program of trapping or monitoring for specific insects. Vineyards, especially those with permanent cover crops or other nearby insectary plantings, provide habitat for an amazing array of insects. Most of these insects are harmless or downright beneficial. It's amazing to drive the mower through the lush vineyard grasses in the spring and become ensconced in a veritable cloud of flying and jumping insects of all sorts. It's a good sign, really, of a healthy, biologically diverse habitat.

Recently, however, invasions of two nonnative grapevine pests into the region have caused sufficient alarm for action by local, state, and federal agencies as well as winegrowers themselves. In 2007 the light brown apple moth, native to Australia, was discovered in California for the first time. A threat to hundreds of crops and ornamental plants, the moth findings triggered federal and state quarantines covering the inspection and movement of affected crops and plants, including grapevines and grapes. Although there was some controversy regarding the potential threat posed by the moth, growers in Sonoma and other affected winegrowing regions moved quickly to mobilize trapping programs to monitor the infestation and treat affected vineyards to thwart the potential threat. Because hundreds of plant species in the area are potential hosts for the light brown apple moth, the prospects for eradication seem pretty slim, so the focus shifted primarily to containment.

Just as we were getting our minds wrapped around the light brown apple moth threat, it was completely blown off the radar when it was discovered that a crop in a ten-acre Napa Valley vineyard was completely destroyed in the fall of 2009 by an infestation of the European grapevine moth, a pest never before seen in California. Unlike the light brown apple moth, the European grapevine moth lives and feeds on basically one plant: grapevines. Producing three to four generations of offspring in a single growing season, the moth is capable of going from near-zero to epidemic proportions in no time at all, as evidenced by the damage in the first infected vineyard in Napa in 2009. In 2010, more than one hundred thousand moths were trapped in Napa Valley vineyards as local growers and local, state, and federal agricultural agencies jumped into action. Also that year, some fifty-nine moths were trapped in neighboring Sonoma County, which was also on high alert as a result of the Napa situation. Approximately forty thousand European grapevine moth traps were placed and monitored throughout affected areas of California in 2010, and aggressive treatment programs were initiated wherever moths were found. Large areas of Napa, Sonoma, and other winegrowing regions were quarantined in 2010, 2011, and 2012, restricting the movement of grapes within and in or out of the quarantined areas. Fortunately, it was still possible to get our grapes to the wineries, but the quarantines required extra work on the part of all involved to insure the movement of grapes from vineyards to wineries did not exacerbate the infestation.

Over the 2010–2011 winter, the Sonoma County agricultural commissioner notified me that a moth had been trapped within five hundred meters of my vineyard; hence, I would be required to implement a precautionary treatment program for all vineyard blocks within five hundred meters of the trapped moth. How did I get so lucky? Less than sixty moths scattered over the sixty thousand acres of vineyards of Sonoma County, and one of them has to be in my neighbor's vineyard!

The first thing we had to do was place pheromone lures throughout our vineyard, which would disrupt any mating activities. The idea was to make the entire vineyard smell like a female moth, thereby disrupting the male moth's ability to find any females to mate with. This entailed hanging several hundred twist ties impregnated with the pheromone throughout the vineyard. We were also required to undergo a spray program timed to coincide with key reproductive milestones of the moths during the season, on the chance that some couples made it through the pheromone cloud and actually did the deed. Fortunately, there were treatments approved for organic production that did not involve the use of deadly, broad-spectrum insecticides. The treatment of choice for organic growers involved spraying a bioinsecticide composed of naturally occurring extract from *Bacillus thuringiensis* bacteria. Often referred to simply as Bt, the insecticide specifically targets leaf-eating moth larvae and is toxic to the larvae when ingested.

Naturally this created quite a bit of extra work, not to mention expense, anxiety, and handwringing. Fortunately, the program seems to be working. Only 9 moths were found in Sonoma in 2011, and only 111 in Napa (down from 100,000 in 2010!). I am optimistic that 2012 will close out my special status in the within-five-hundred-meters club, and I can return to my regularly scheduled programming. Which means staying vigilant in my powdery mildew program and keeping my fingers crossed that no other pests come my way.

6

SOME THINGS GET BETTER WITH AGE

Vitis vinifera evolved over millions of years to thrive in a particular climate and ecosystem. This environment did not include the fungus *Erysiphe necator*, which we know today as the cause of grape powdery mildew. The evolutionary path for *Vitis vinifera* was similarly devoid of contact with the fungi that cause downy mildew and black rot. The consequences of this particular evolutionary path have become abundantly clear since the introduction of the powdery mildew, downy mildew, and black rot fungi to Europe in the nineteenth century.

Of course, there is no assurance that *Vitis vinifera* would have developed a resistance to powdery mildew had the two evolved together. But the odds are certainly good that the susceptibility of *Vitis vinifera* to powdery mildew infection would be much different from what we find today. Indeed, such is the case for other species of grapevines that are native to North America.

Of the sixty or so known species of Vitis grapevines around the world, more than half are native to North America. It was from these North American grapevines that the cure for the phylloxera epidemic was found. Today, rootstocks from **Vitis berlandieri**, **Vitis riparia**, and

Vitis rupestris form the basis of the phylloxera-resistant rootstocks that the wine world depends upon. Why wouldn't a similar strategy work for powdery mildew?

Many native North American grapevines have shown resistance to powdery mildew infection, as well as to downy mildew and black rot, but to date this resistance has not proven practical for transference to *Vitis vinifera*. The use of North American rootstocks confers no resistance, as it did amply in the widespread use of such rootstocks to battle phylloxera. Crossbreeding of *Vitis vinifera* with North American varieties might be another route to resistance, but such efforts have not produced wine of the same quality. Indeed, varieties such as Rubired and Royalty, which are hybrid crosses of *Vitis vinifera* and *Vitis rupestris* cultivars, have shown moderate resistance to powdery mildew. But the wines from these hybrids are not of the same quality as that of their *Vitis vinifera* parents. Disease-resistant hybrids are fairly common in the eastern United States, where disease pressure threatens the viability of *Vitis vinifera* vineyards, but they are quite rare in California and most other premium winegrowing regions of the world.

There are other mechanisms besides coevolution by which plants might acquire resistance to pathogens that are not native to their environment. For example, resistance mechanisms evolved to fight one pathogen might coincidentally provide resistance against another, similar nonnative pathogen. Also, abiotic or environmental factors might cause adaptations in a plant that serendipitously make the plant less susceptible to a particular pathogen. While *Vitis vinifera* is notoriously susceptible to powdery mildew, we do know that this susceptibility is limited to certain stages of growth in the vines and fruit.

ONTOGENIC RESISTANCE

We have long known that early in the growing season, young shoots, leaves, and fruit are very susceptible to infection. By late summer, however, mature leaves and canes are no longer susceptible to infection, and mildew colonies are more virulent in their sporulation on young leaves than on mature leaves. Most growers have seen this firsthand, as the vines continue to produce new leaves into July or August. In such cases, the mature leaves growing along the primary branch of the cane or shoot are healthy and green in midsummer, but young leaves growing on secondary branches in the shady interior of the canopy yellow and wither with mildew. Not only are these young leaves on secondary shoots more susceptible to infection than the mature leaves on the main shoot stem, but they are more difficult to reach with fungicides. The thick canopy of midsummer both provides the shade mildew thrives in and also serves to block penetration of protective fungicide sprays.

In a similar vein, it has been believed for several decades that individual grape berries are uniformly susceptible to new infection up to a ripeness level of eight to twelve degrees Brix (a measure of sugar content in the grapes), which coincides with the onset of veraison. In common practice this often translates into regular, periodic fungicide sprays for more than four months, from bud break to veraison.

In both cases, with leaves and fruit, we see that something changes during the growing season. The same leaf that is a virtual mildew magnet in April or May miraculously becomes impenetrable only a few weeks later. What changes during the season to confer this new immunity? When does it happen?

In recent years research has shown that resistance in the fruit to powdery mildew infection actually begins much earlier than veraison, the practical implications of which could be huge for more effective mildew control and reduced pesticide usage. Researchers at Cornell University and elsewhere have shown that resistance in the young grape berries begins to develop immediately after fruit set, with the berries going from highly susceptible to nearly immune in less than thirty days. This contrasts with the widely held view of the development of resistance at veraison, when sugars in the ripening grapes have accumulated to around eight to twelve degrees Brix, which typically occurs about two months after bloom.

Looking more closely at this phenomenon provides a picture of rapidly developing resistance over a period of several weeks. Resistance to infection has been shown to decrease from basically zero at thirteen days after bloom to about 80 percent at twenty days and 95 percent at thirty days, reaching better than 99 percent resistance by fifty days after bloom. This timeline for the development of resistance has been shown to apply equally well to both the fruit and the rachis (stem) of the grape cluster. This type of disease resistance, which develops over time according to the age or developmental status of the plant, is called ontogenic resistance. Similar timelines have been shown for the development of ontogenic resistance to downy mildew and black rot.

Understanding the course of development for ontogenic resistance can be a powerful tool for more effective and efficient disease control in the vineyard. Unfortunately, the development of ontogenic resistance varies with different climatic conditions, which makes it much more difficult to define the timeline for vineyards across different growing regions and climates. One important factor is the effect of climate on bloom in the vineyards. In cold climates, where winter temperatures are frequently below freezing, the period of bloom for an entire vineyard may be as short as two days, whereas in the milder winter climates, where most wine grapes are grown, bloom can take up to two or even three weeks to complete, as is generally the case here in Sonoma. This means that the youngest berries in a warm-climate vineyard may develop ontogenic resistance as much as two to three weeks later than the oldest berries in the vineyard. On a single vine, even within a single grape cluster, there may be berries that differ in this measure of maturity by two or three weeks.

The development of ontogenic resistance for the leaves takes a similar course, with young leaves being most susceptible to powdery mildew infection at about five to six days after unfolding, with

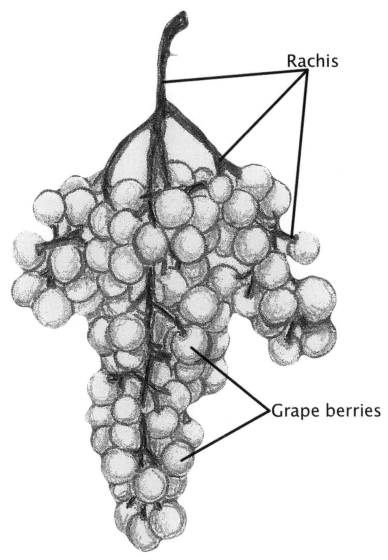

Main features of a grape cluster. (Drawing by Sherry Rahn.)

susceptibility declining to near zero by the time the leaves are two to three weeks old. Unlike grape clusters, vines produce new leaves fairly continuously from bud break until veraison, though the amount of new leaf area produced by the vine peaks around bloom and declines thereafter as the vine devotes more energy to fruit development than vegetative growth.

The susceptibility of leaves to infection is one of the most important factors in determining the disease pressure the young grape flowers and berries face at the critical period when disease susceptibility in the emerging fruit is high. The prevalence of disease on the leaves just before and after bloom has been shown to be critical in determining the incidence and severity of mildew on the fruit at maturity. This underscores the importance of managing disease pressure in the period leading up to and just after bloom, as the level of disease on the leaves at the time of maximum fruit susceptibility will be the most important determinant of fruit infection later in the season. This period of heightened susceptibility just before and around bloom applies not only to powdery mildew but also to downy mildew, black rot, and botrytis infections. Awareness of and response to infection events that occur during critical prebloom periods is key, particularly (1) infection events such as rainfall (e.g., ascospore release) in the first month after bud break when the leaves around the clusters are most susceptible, and (2) infection events in the month leading up to bloom, when susceptible leaf area is at a maximum, and increased canopy size makes spray coverage more challenging, especially for organic growers using contact fungicides.

For all the good that comes of ontogenic resistance to powdery mildew in grapevines, its biological origins and mechanisms remain somewhat mysterious. Is it a passive mechanism, brought on by changes to the plant tissues that occur naturally with age? For example, to what extent do increases in leaf cuticle thickness and cell wall lignification that occur as the leaf ages play a role in preventing penetration of the leaf surface by the mildew? Alternatively, is it an active response by the plant to the presence of a mildew attack, such as the production of enzymes that attack or prevent penetration and growth of mildew tissues in the plant?

The answers to these questions remain unclear, though a great deal of research has been underway for much of the past decade searching for them. What we do know is that as the leaves and fruit mature, it becomes increasingly difficult for the *Erysiphe necator* fungi to colonize these vine tissues.

Initially, as ontogenic resistance begins to develop after fruit set, mildew spores are still able to germinate on the berry or leaf surface. However, the mildew is unable to penetrate the berry or leaf surface to establish the roots through which the fungi feed on the vine nutrients. More importantly, the fungi are unable to grow and reproduce. Unable to penetrate the leaves or berries and draw the nutrients needed to sustain growth and reproduction, the mildew colony eventually dies. As the development of ontogenic resistance nears completion, mildew spores are no longer capable of germinating on the leaf or berry surfaces, and the surface area of the leaves or berries colonized by earlier infections ceases to grow. Consequently, the real damage from powdery mildew results from infections that occur early in the season, before the onset of ontogenic resistance. Although ontogenic resistance may eventually halt the growth of these early infections, the damage will have already been done.

The development of ontogenic resistance during each growing season is a blessing for growers, particularly given the high level of susceptibility that precedes its onset. Given the struggle we face each season to get to the point where powdery mildew is no longer an issue for that season, it would only be that much worse if the threat persisted all the way to harvest.

7

THE WINEGROWER'S CHALLENGE

O ne of the great rewards of farming is the wondrous sense of renewal that comes with the arrival of each new season. By season, I don't mean winter, spring, summer, or fall. I am referring to the new growing season that comes each year. It just so happens that this coincides roughly with the vernal equinox, which is the official beginning of spring. Around this time the vines awaken from their winter dormancy, and the cycle of growth starts anew.

The winter is a dormant period of sorts for the winegrower as well. Apart from pruning, which takes place when the vines are dormant in the winter, it is a time of rest, recreation, and reflection. Naturally, much of the reflection is cast upon the lessons of the prior season. No matter how long one goes about a particular seasonal profession or avocation, the most salient lessons are those of the most recent season. Winegrowing is no different. We enter each new season determined to avoid the afflictions that pained us most in the prior year. Never mind that bigger, more persistent problems may still linger from seasons in the more distant past.

I have found that after a bad or persistent mildew season, I spend a good deal of time trying to figure out what I might have done differently.

Were there signals or warning signs I missed? What should I do differently this year? Discussions with other growers and the data I keep detailing all of our vineyard inputs and activities provide a lot of insight into what happened and why. As for what I might do differently, many of the choices are constrained by our farming philosophy—or, more specifically, the options allowed by or consistent with that philosophy.

Mildew control, however important, is just one part of a bigger picture. Many of the practices and materials we use will be governed by the farming philosophy we employ. A grower employing conventional farming practices, for example, will have a very different approach and set of tools at his disposal than an organic farmer.

What I call a farming philosophy others might call a farming system. In cases where a formal certification system governs the farming practices, as with a certified organic farm, "farming system" might seem the more appropriate term. However, even then the rules and regulations governing certified organic farms do not specify all the practices the farmer might employ. For example, the US Department of Agriculture rules that govern certified organic farms say nothing about tilling versus no-till farming, other than specifying the latter as "practices that maintain or improve the physical, chemical, and biological condition of soil and minimize soil erosion." The decision to till or not then falls under a broader rubric, which I believe the word "philosophy" more accurately captures.

Nonetheless, in modern viticulture there are arguably four main farming philosophies or systems within which almost every vineyard operation fits. The boundaries between these philosophies are not always clear-cut, but they are easily distinguished in fairly broad strokes. In some cases, as with certified organic and biodynamic farms, many of the boundaries are very clearly defined in the rules and regulations that govern those farms. The four main farming philosophies used in vineyards around the world today are conventional, sustainable, organic, and biodynamic. The distinctions among these alternatives are interesting and sometimes controversial.

MODERN FARMING PHILOSOPHIES

Conventional Farming

When it comes to modern viticulture, "conventional farming" has almost become a pejorative label. I don't personally know anyone who would say they farm conventionally. In fact, adherents of the more progressive farming philosophies might use this label to describe their polar opposites. What is this object of scorn, and why is it so vilified?

The term "conventional farming" and its synonym "conventional agriculture" sound innocuous enough. But not since the counterculture revolution of the 1960s has the word "conventional" carried such a derogatory meaning. Oddly enough, in a big picture or historical view of agriculture, there is nothing conventional about conventional agriculture. Conventional agriculture, as we know it today, arose in the twentieth century with the industrialization of agriculture, enabled by the development of synthetic fertilizers and pesticides along with new machines that brought unprecedented advances in the scale and productivity of farming operations.

For thousands of years agriculture had developed using natural, organic materials for fertilizer, such as sewage, animal manures, bones, and other remains, along with crop rotation and green manures (cover crops). Although the mining of mineral nutrients became important for Western agriculture in the nineteenth century, the defining event for conventional agriculture was the invention of an industrial process for synthesizing nitrogen fertilizers from atmospheric nitrogen. The German scientist Fritz Haber developed the process in 1909, for which he was awarded the Nobel Prize in chemistry in 1918.

The development of industrial agriculture led to the agricultural philosophy within which conventional agriculture fits today. Conventional agricultural practices need not be on an industrial scale, but the basic precepts of conventional agriculture are descended directly from industrial agriculture. The most basic precept is to leverage industrial technology to maximize agricultural output, productivity, and efficiency. Apart from machines, the most important technologies are agrichemical fertilizers, fungicides, insecticides, and herbicides. These are synthetic chemicals developed and manufactured for agricultural purposes by industrial agrichemical companies.

In conventional agriculture, synthetic fertilizers are used primarily to maximize crop yield. While fertilizers are used in all types of farming systems, what sets the conventional philosophy apart is its focus on feeding the crop rather than the soil and supplying nutrients at levels that will maximize crop yield. Conventional fertilizers did revolutionize farm productivity in the twentieth century, dramatically increasing crop yields while lowering the cost to consumers of many food products. Indeed, many supporters of conventional agriculture claim that it would not be possible to feed the world's large and growing population without these fertilizer practices. While the latter claim is arguable, there is no doubt that sustainability and pollution issues associated with these fertilization practices offset many of the benefits. For example, groundwater pollution from fertilizer runoff is a major source of water pollution, contributing to dead zones such as the one in the Gulf of Mexico off the coasts of Louisiana and Texas.

Pesticides in conventional farming, especially fungicides and insecticides, are applied prophylactically to eliminate potential disease or insect problems before they occur and to treat observed infestations. While there is certainly prophylactic pesticide usage in other farming philosophies, it is generally discouraged. In many ways, what sets conventional farming apart are the materials used and the frequency or quantity of usage. One major distinction is the reliance on synthetic chemicals, most of which are systemic, which means they are absorbed and translocated within the plant. Most of the pesticide residues found in food products are the result of systemic pesticide use. As we shall see, while the sustainable farmer may use many of the same materials as the conventional farmer, how those materials are used often differs.

Similarly, herbicides are used programmatically to eliminate any and all noncrop vegetative growth that might compete with the crop for water, nutrients, or sunlight. The conventional farmer seeks to eliminate noncrop vegetation. In contrast, many biodynamic and organic growers, and even some sustainable growers, see noncrop vegetation as part of a healthy and diverse ecosystem. In these philosophies, noncrop vegetation is tolerated if not encouraged. Noncrop vegetation is managed rather than eliminated. This is particularly true in permanent crops like vineyards and orchards. Competitive vegetation is less tolerable in low-growing and shallow-rooted row crops like vegetables, but even here the biodynamic and organic growers eschew chemical herbicides.

A notable example that illustrates the hidden costs of the conventional approach to pesticide use is the case of genetically engineered (GE), pesticide-resistant crops. The idea behind these crops is to enable farmers to reduce tillage, which is good for the soil, reduces air pollution, and increases productivity by reducing weed competition. The unintended consequences, though, have been dramatic increases in pesticide use and the development of pesticide-resistant weeds. This is a vicious circle, as the resistant weeds require ever-greater volumes of pesticide for control. Contamination of non-GE crops through cross-pollination and the spread of the pesticide-resistant weeds to neighboring farms are also cause for alarm.

The common thread in the use of agrichemicals in conventional farming is the emphasis on yield and productivity. In commodity farming—growing soybeans or corn, for instance—maximizing crop yield may be critical to economic viability. For specialty crops, like wine grapes or even peaches destined for farmers markets, maximizing crop yield is likely secondary to other factors, such as fruit quality. Hence, conventional farming is much more prevalent among commodity crops than it is for wine grapes. Conventional winegrowing, as defined here, is pretty rare in the premium winegrowing regions of California. The quality criteria used by wineries and growers for premium wine grapes are not consistent with a conventional farming focus on maximum yields. However, wine grapes destined for inexpensive jug wines and other wines on the low end of the price spectrum may be a different matter all together. For example, the average price paid per ton of grapes by wineries for grapes grown in Sonoma County in 2011 was $2,081 per ton. In contrast, in the Fresno, California, area, a major farming region for grapes used in inexpensive wines, the average price in 2011 was $346 per ton. To produce the same revenue per acre of grapes, the Fresno grower needs to produce six times the yield per acre as the Sonoma grower.

Mildew control in conventional viticulture today relies primarily on systemic, synthetic fungicides developed by modern industrial agrichemical companies. These fungicides are very powerful and effective against powdery mildew and other fungal pathogens that infect grapevines. The benefits are that they can be used in relatively small quantities and do not need to be applied as frequently as the organic alternatives. Fewer applications also mean fewer tractor passes through the vineyard and a lower carbon footprint. On the downside, these are very powerful chemicals with unknown environmental effects. The fact that they are systemic and persist within vine and fruit tissues may also be problematic, as they end up in the foods and wines we consume. An important but lesser-known disadvantage is that the mildew fungi develop resistance to these fungicides with repeated exposures. So it is very important for growers to use these fungicides selectively and to follow prescribed rotation schedules using different classes of fungicides to prevent development of fungicide-resistant strains of mildew.

Sustainable Farming

Sustainable winegrowing has become the mantra of many farmers and winegrowing regions around the world. While the promotion of sustainable agriculture has been criticized by some as merely "green washing," in recent years there has been a great deal of progress in codifying sustainable practices and the development of certification programs. Organizations promoting sustainable winegrowing exist in the United States, France, South Africa, Australia, New Zealand, and other major winegrowing countries.

The largest formal program for sustainable winegrowing in the United States is run by the California Sustainable Winegrowing Alliance (CSWA), which was founded in 2003 to promote sustainable practices in the wine industry. The CSWA defines sustainable winegrowing as growing wine in a way that's environmentally sound, socially equitable, and economically feasible. The CSWA runs the Sustainable Winegrowing Program, translating this high-sounding mission into discrete actions for winegrowers around the state. The program consists of a repeatable cycle of self-assessment workshops, followed by detailed performance reports and action plans, to implement improvements in sustainability in vineyard and winery operations. Since its inception, more than twelve hundred vineyard organizations have participated in its workshops, representing nearly 50 percent of the total vineyard acreage in California.

The Sustainable Winegrowing Program assesses a broad spectrum of vineyard practices such as viticulture, soil management, vineyard water management, pest management, and ecosystem management. The soil-management assessment, for example, includes fifteen criteria for assessing the management of soil and plant nutrition, soil fertility, and tilth, erosion, and the use of cover crops. A few years ago I participated in one of the Sustainable Winegrowing Program workshops and found the self-assessment to be very thorough and the resulting analysis of my vineyard operations to be both insightful and helpful.

An important initiative recently added to the Sustainable Winegrowing Program is the Certified California Sustainable Winegrowing program for certification. Modeled after certification programs for organic production, the program provides for independent, third-party auditing of vineyard sustainability practices and action plans for continuous improvement. Launched in 2010 the program certified nearly sixty thousand acres of grapes, or about 11 percent of the state-wide total, in its first two years.

There are similar sustainability initiatives taking place at the local level in California and other winegrowing regions. In California, programs such as the Central Coast Vineyard Team Sustainability in Practice Certification and Lodi Certified Green also provide certification for sustainable vineyards, but more importantly these and others provide local support and resources to support sustainable winegrowing. Another model sustainability program is Oregon's LIVE, where LIVE stands for Low Input Viticulture and Enology. The Oregon LIVE program advocates limiting the amount of raw materials used in vineyard and winery production, focusing on materials such as pesticides, fertilizers, water, chemicals, and fuel.

There is no doubt that the sustainability movement has been picking up momentum in the past decade, and the broad adoption of sustainable practices is improving the farming community's stewardship of the land being used for growing wine. However, this has not eliminated the disagreements over the pros and cons of different farming philosophies.

To some, sustainable farming is seen as the middle ground on a continuum, with conventional farming on one extreme and the more enlightened or progressive organic and/or biodynamic farming on the other extreme. As it pertains to the use of chemicals versus no chemicals in the application of fertilizers and pesticides, there is some logic to this simple linear continuum, but at the same time it dramatically oversimplifies the distinctions.

It is true that all of the major sustainable winegrowing programs permit the use of the same types of agrichemicals used in conventional farming, whereas neither the organic nor biodynamic philosophies allow such materials. But the use of agrichemicals in sustainable agriculture is much more nuanced than in conventional farming. For example, in sustainable farming, fertilizers are ideally applied only to address deficiencies based on soil and plant tissue analysis. In contrast, conventional farming typically applies fertilizer in a formulaic manner on a seasonal or calendrical basis. This often leads to excessive applications of fertilizer, contributing to groundwater pollution and increased pesticide usage to counter disease, insect, and weed problems that accompany excess vigor and fertilization.

Sustainable farming advocates would also point out, as would proponents of conventional agriculture, that the chemicals versus no chemicals distinction overlooks the fact that some pesticides approved for organic production are quite toxic, even if they are naturally occurring elements of the earth. In this view, fungicides approved for organic production, such as sulfur and copper, may be natural but are toxic to a broad range of organisms. Advocates of sustainable and conventional farming argue that spraying five to ten pounds of these toxic materials per acre is no better than spraying a few ounces of a synthetic chemical fungicide, and the increased frequency of spraying that comes with organic pesticides also contributes to an increased carbon footprint. Similarly, many sustainable farming advocates argue that a much smaller carbon footprint comes with one or two applications of herbicide per year and is more environmentally sustainable than the repeated tractor passes needed during the season to control weeds under the vines using mechanical tillage in lieu of herbicides. Whether or not these types of arguments are enough to make you comfortable with the use of synthetic agrichemicals, it is clear that there is not a linear continuum of sustainability or environmental sensitivity of the sort described above.

Clearly, for advocates of organic farming, all of these sustainable winegrowing programs permit the use of chemicals that are anathema to the organic farmer. While allowing the use of materials prohibited in organic or biodynamic farming, the sustainable programs do support the use of organically approved fertilizers and pesticides. At the same time, the sustainable winegrowing programs cover important topics that are not explicitly addressed by the organic or biodynamic standards, such as promoting minimal soil tillage, energy efficiency, and water conservation.

I strongly support the sustainable winegrowing movement, as it is driving positive changes to the way wine grapes are grown, including the use of fewer and less toxic pesticides. I find the winegrowing community as a whole to be responsible and environmentally conscious, and most vineyards that I am familiar with are family farms whose owners want nothing more than to pass on their vineyards to their children or grandchildren. Sustainability means the next generation inherits a farm that is more healthy and productive than the one their parents began with.

Organic Farming

Organic farming has become something of a lighting rod issue in some circles. You have one side claiming organic foods are healthier than conventionally grown foods, while their opponents charge that organic foods are overpriced and no different from other foods. These debates are flamed by the occasional charges of fraud among farmers for passing off conventional foods as organic or violating organic rules by cutting corners in their production methods. Similarly, charges of fraud have been levied against a few organic fertilizer producers. Most of these issues play out around organic produce sold in farmers markets and supermarkets across the country. Organic fruits and vegetables command a premium, and anywhere money is to be made, such problems seem to follow.

Unlike with produce, there seems to be no pricing premium for organically grown wine grapes. Granted, a few wineries seek or specialize in wines made from organic grapes and may offer incentives to growers to convert their production to organic, but as a whole the organic premium that exists for other foodstuffs does not exist for wine grapes. Interestingly, it seems that the drivers behind organic wine grape production are mostly philosophical rather than economic.

However deeply held one's philosophical convictions may be, the use of the word "organic" to describe any agricultural product in marketing literature or product packaging is tightly regulated. This is true in the United States and many other countries as well. In the United States, the US Department of Agriculture (USDA) National Organic Program (NOP) governs organic production and labeling. As applied to wine, there are separate standards governing the growing of organic wine grapes and the production of organic wine. Unfortunately, this distinction can be quite confusing for consumers, and I will attempt to clarify it before moving onto the details of organic farming.

It turns out that very little organic wine is produced in the United States, even though there are a great many organically grown grapes. The reason is the additional restrictions that the USDA NOP places on wine to be labeled as organic. First and foremost, organic wine must be made from certified organic grapes. However, in turning grapes into wine, winemakers often add other ingredients to facilitate or manage the fermentation process. Most commonly these include yeasts for alcoholic fermentation, bacteria for malolactic or secondary fermentation, nutrients to support these yeasts and bacteria, and sulfur dioxide or sulfites to preserve the wine against oxidation and microbial spoilage. However, the NOP standards explicitly exclude the use of sulfur dioxide in wine to be labeled as organic. As sulfur dioxide is critical to the stability and shelf life of wines, relatively few producers are willing to put out a product without any added sulfites. The risk to their products and reputations is simply not worth it.

On the flip side, wines in the United States may be labeled as made with organic grapes if the grapes in the wine are 100 percent certified organic, various other organic production and handling requirements are met, and the total sulfite concentration does not exceed one hundred parts per million. The regulations governing organic wine and wine made with organic grapes restrict the yeasts, nutrients, and other additions to an approved list of natural ingredients, but it is really only the sulfite restrictions that are contentious. The European Union, for instance, has similar distinctions between wine made with organic grapes and organic wine but allows both categories to contain limited amounts of sulfites.

Back to farming. . . The USDA NOP has three major sets of requirements that impact organic winegrowing and its operations: (1) production and handling requirements that govern farming practices and allowed materials, (2) documentation requirements that include detailed record keeping and a formal organic systems plan, and (3) the requirements that govern the formal organic certification process.

The NOP organic production standards govern a broad spectrum of farming practices. They specify the buffers that must exist between organic and nonorganic fields, the allowable soil fertility and crop-nutrition practices, the types of seeds and planting stock that may be used, crop-rotation practices, and pest- and disease-management practices. In addition, there are official lists of approved products that may be used as farming inputs, such as fertilizers or pesticides. The most widely used such list of products is produced by the nonprofit Organic Materials Review Institute (OMRI).

The NOP farming regulations can be quite specific. For example, they specify that compost used for plant nutrition and soil fertility must have a certain carbon-to-nitrogen ratio and be produced in a process conforming to prescribed durations and temperatures. Overall the NOP organic production requirements promote sustainable farming practices using natural materials, severely limit the use of synthetic or toxic materials, and prohibit use of or contact with genetically modified organisms.

In addition to the regulation of allowable and prohibited practices and materials, the NOP requires detailed record keeping of all production, harvesting, and product-handling operations. These records must be kept for at least five years and be made available upon request for inspection or audit. In addition the grower must maintain and update annually a detailed organic production plan, which includes the following:

1. A description of farming practices and procedures, including their frequency
2. A list of all substances to be used as a farming inputs, including their composition and source, location(s) where they will be used, and documentation of commercial availability
3. A description of monitoring practices and procedures and the frequency with which they will be performed
4. A description of the record-keeping system for production, harvesting, and product-handling operations
5. A description of the practices and barriers used to prevent commingling of organic and nonorganic materials or produce and to prevent contact of organic production and handling operations and products with prohibited substances

Finally, to be certified requires documented compliance with the NOP rules and regulations, including the annually updated organic production plan and record-keeping documentation described above, which are provided along with applicable fees to an approved certifying agent. One must also permit inspections of all records, farming operations, and facilities by the certifying agent. One of the largest and oldest certifying agents in the country is the California Certified Organic Farmers (CCOF). As of the end of 2011, the CCOF has twelve thousand acres of certified organic vineyards in California, or about 2 percent of the vineyard acreage in the state.

Certified Organic represents a relatively small slice of the total vineyard acreage and understates the widespread adherence to organic farming as a philosophy among winegrowers, particularly in the premium winegrowing areas. There are several explanations for this discrepancy. First, there is relatively little economic incentive for growers to incur the costs of organic certification, as there is no market premium for organic grapes, and the market for organic wines and wines made from organic grapes is relatively small. Second, many growers practice organic farming because they see it as the right thing to do but do not wish to incur the additional costs, restrictions, and overheads associated with certification.

Given all this controversy and regulation, it seems reasonable to ask, Just what is organic farming? What I have described above mostly pertains to the formal organic farming system as regulated by the USDA, and one can find similar rules and regulations in other countries as well. These formal organic farming systems embody or codify broad organic farming philosophies. Some tenets of the organic philosophy are inclusive, by which I mean to say that they embrace certain types of materials and practices, and other tenets are exclusive in the prohibition of specific types of materials and practices. In the broad public view, organic farming is often associated primarily with what it prohibits, namely, (synthetic) chemical fertilizers and pesticides.

But as an organic farmer, I find the most important tenets of organic farming to be the broader guiding principles, in which the farm is viewed as a highly interdependent ecosystem of living organisms, both above and below the ground. The crops grown by the farmer may be most important economically, but the environmental and social sustainability of the farm depends on the health of the entire ecosystem of the farm and its surroundings. It is this view of farming as ecology that is the core value of organic farming and is reflected in the use of the term "biological" or "ecological" rather than "organic" in other countries. For example, organic farming is called *agricultura ecológica* in Spain, *biologische landwirtschaft* in Germany, *l'agricultura biologica* in Italy, and *l'agriculture biologique* in France. This core principle of the organic philosophy celebrates ecology and biodiversity and farming practices that promote biodiversity in healthy farm ecosystems.

This is where one can run into the tension between organic farming as a philosophy and organic farming as a system. While most or all of the formal organic farming systems, such as the USDA NOP, embrace this ecological view of farming, the implementation of the systems does enable organic farming by the letter of the law rather than the spirit. It is possible to apply a conventional farming mentality to organic farming, substituting fertilizers and pesticides approved for organic production for conventional alternatives. The letter of the law in the NOP discourages this, but it no doubt occurs and is the basis for much criticism of the industrialization of organic farming and the rise of large-scale organic farms.

This ecological view of farming drives the core values and practices of organic farming. For example, an ecological approach to fertilization in organic farming will focus on feeding the soil to foster a rich soil food web that will in turn provide all the nutrients the plants need. This contrasts with the conventional view of fertilization, which is to feed the plants, where the soil is just the medium in which these nutrients are delivered. I was reminded of this distinction recently when attending a presentation on soil fertility and vine nutrition by a local expert, a soils scientist by training. I realized that the speaker was basically a soil chemist and viewed fertilizers primarily as chemical inputs to the soil that were either available or unavailable for uptake by the plants, depending on their chemical properties and the chemical properties of the soil. Not one word about soil biology or the role of soil organisms in the cycling of nutrients and availability of nutrients for plant uptake. A presentation on soil fertility and vine nutrition by a soils scientist with an organic farming bent, more soil biologist than soil chemist, would have been completely different. One could follow the advice of the soil chemist who delivered the presentation I referenced and operate within the strictures of the NOP, but the spirit at the heart of organic farming would be lost.

Similarly, organic farming promotes biodiversity as a means to maintain a healthy and balanced ecosystem. Biodiversity on the farm promotes natural biological pest controls, which reduces pesticide use. Permanent cover crops, insectaries within the farm boundaries, and buffer zones of native habitat in and around the farm crops all promote natural predators for many types of pests. Natural, biological controls can reduce pest pressure below the economic threshold where pesticide applications are required.

Of course, organic farming is also about the materials used on the farm, most notably fertilizers and pesticides. A common misunderstanding is that organic farms use no pesticides. As noted above, a conscientious organic farmer can avoid pesticide use in many cases, using biological controls and other cultivation practices. Nonetheless, a foundational principle of organic farming is that all farm inputs, be they fertilizers, pesticides, or planting materials, should be natural materials. Fertilizers are derived from plant and animal sources, such as manure and other plant or animal by-products (e.g., compost, fish emulsion, bone meal, blood meal), as well as naturally occurring rocks and minerals. Pesticides are likewise derived from natural sources, including natural elements such as copper and sulfur, along with biological pesticides derived from plants, bacteria, and fungi. Granted, even some natural materials are toxic, and in general conscientious organic farmers will tend toward softer, less toxic materials wherever possible. The flip side of the natural materials requirement is the prohibition against synthetic chemicals, which in most formal organic farming systems extends to genetically modified organisms as well. So, while the common view of organic versus conventional farming focuses on the natural-versus-synthetic distinction, I believe that the most important principles of organic farming are the broader ecological principles that provide the foundation of the organic farming philosophy.

Biodynamic Farming

Whereas organic farming is an ecological approach to farming, biodynamics is both a spiritual and an ecological farming philosophy. While biodynamic farming is properly considered a superset of organic farming, it carries with it certain spiritual or mystical notions held dear by its adherents and often ridiculed by its detractors. So, if organic farming is something of a lightning rod issue in certain circles, then biodynamics might be considered a full-on electrical storm.

The discipline of biodynamic farming is the outgrowth of a set of lectures delivered by the Austrian philosopher and polymath Rudolph Steiner in 1924. Steiner's *Agricultural Course* lectures espoused an agricultural approach that includes most of the basic precepts of what we know today as organic farming, supplemented by practices based on his anthroposophical views of how cosmic forces affect life on earth. As described by Steiner, "Anthroposophy is a road to knowledge leading the spiritual part of the human being to the spirit of the universe." So Steiner infused his farming philosophy with practices he divined to align the forces of the cosmos with key events and activities on the farm. The most prominent are planting, cultivating, and harvesting prescriptions that are based on a lunar calendar, although Steiner also attributes considerable influences to other planets in our solar system as well as to more distant bodies in what he calls "the most distant cosmic forces."

The Demeter cooperative was formed in 1928 to codify Steiner's farming principles into a formal program, which became the Demeter Biodynamic Farm Standard. Today, Demeter International is the parent organization for a network of organizations in forty-five countries that govern biodynamic farming practices and certification. Biodynamic is a registered trademark of the Demeter Association in the United States, and use of the term as applied to a farm or any farm product is subject to certification by the Demeter organization.

A central premise of biodynamic farming is the view of the farm as a holistic, living organism, in which all of its many facets are intricately intertwined, including the soil, plants, insects, birds, rodents, and all its other forms of life. This is the ecological foundation biodynamics shares with organic farming, though the Biodynamic Farm Standard goes further than organic standards such as the NOP's in prescribing practices that promote biodiversity and a holistic approach to the farm as an ecosystem. For example, the Biodynamic Farm Standard frowns upon practices such as planting crops while at the same time using pesticides to eliminate insects, weeds, and other forms of life seen as injurious or competitive to those crops, as the pesticides disrupt the balance and life forces needed to nourish and sustain the farm and its crops.

Similarly, the nourishing of the soil and recycling of nutrients with organic matter such as cover crops, manure, and compost, along with minerals and biodynamic preparations, are fundamental biodynamic practices. The biodynamic ideal sees the farm as a closed system, with all of these nutrients produced within the ecosystem of the farm itself. The practical realities of biodynamic certification recognize the difficulty of achieving this ideal and do permit the importation of nutritional materials from off the farm. As far as winegrowing is concerned, the limitations on imported nutrients are rarely an issue as the limits are generally well above what would likely be applied, for premium wine grapes anyway. For example, the annual limit for imported nitrogen fertilizer application of eighty-six pounds of nitrogen per acre is well above accepted or advisable winegrowing practices.

Biodynamic disease and insect control is ideally managed using biological controls that are encouraged through the maintenance of habitat diversity, such as with crop rotation, cover crops, insectaries, and hedgerows. Managing crop vigor with balanced nutrition and irrigation and attention to light penetration and airflow are also important elements of disease and insect control. Pesticides approved for organic production are meant to be a solution of last resort, but the realities of winegrowing dictate the regular use of fungicides to control powdery mildew, as well as to control downy mildew, botrytis, and other common fungal pathogens in many climates.

Biodynamic weed control employs the same basic controls as organic farming, relying primarily on mechanical controls such as mowing, plowing, and grazing. The biodynamic standards also emphasize prevention or exclusion through the timing of plantings and irrigation, mulching, and avoiding the spread of invasive weeds by applying knowledge of the weed life cycle and nutritional practices that discourage weed growth.

The biodynamic farm standard also encourages, but does not require, the integration of livestock into the farm ecosystem for manure production, nutrient recycling, and biodiversity. The standard also specifies that 10 percent of the total farm acreage be set aside as a biodiversity reserve, meaning no agricultural activities are conducted on that land.

Biodynamic certification is not unlike that for USDA organic certification, requiring detailed record keeping and documentation, inspection by an official certifying entity, and annual renewal of the certification. In the United States, the USDA NOP standards provide a baseline set of requirements for biodynamic certification. As with organic certification, biodynamic certification relies on the same official lists of materials approved for organic production. Materials and products used in biodynamic production in the United States must be approved for organic production by one of the USDA-sanctioned organizations, such as the OMRI.

Some proponents of biodynamics position it as "beyond organic," with the often not-so-subtle implication that biodynamic is "more organic" than organic or more capable of expressing the maximum potential of a particular vineyard. Since biodynamic farming is literally a superset of organic farming, the believability of these claims comes down to just how much stock one places in the spiritual facets of biodynamics that take it beyond what we typically think of as organic farming. While there is little or no empirical evidence in the agricultural research realm to differentiate the benefits of organic from biodynamic farming, spiritual or mystical claims rarely show well in the harsh light of science. I personally am rather skeptical of the mystical foundation of biodynamics, beyond its organic farming foundation. However, there are clearly many converts and believers, with many testimonials to its efficacy.

Biodynamic Preparations

Among the most controversial tenets of biodynamic farming are the nine biodynamic preparations. These preparations are made from herbs, minerals, and manures and are applied to the soil or sprayed directly on the plants in minute quantities. These preparations are akin to homeopathic medicines and are intended to concentrate and channel cosmic forces that act on plant growth. Although much of the backstory surrounding the preparations is not described in the Biodynamic Farm Standard, Steiner's original *Agricultural Course* lectures provide rather interesting insights into the motivations for these practices. Here is Steiner describing what came to be known as Preparation 500 in Lecture 4 of the *Agricultural Course:*

> We take manure, such as we have available. We stuff it into the horn of a cow, and bury the horn a certain depth into the earth—say about 18 in. to 2 ft. 6 in., provided the soil below is not too clayey or too sandy. (We can choose a good soil for the purpose. It should not be too sandy). You see, by burying the horn with its filling of manure, we preserve in the horn the forces it was accustomed to exert within the cow itself, namely the property of raying back whatever is life-giving and astral. Through the fact that it is outwardly surrounded by the earth, all the radiations that tend to etherealise and astralise are poured into the inner hollow of the horn. And the manure inside the horn is inwardly quickened with these forces, which thus gather up and attract from the surrounding earth all that is ethereal and life-giving.

Preparation 500 is made from cow manure that is fermented in a cow horn that is buried in the soil for six months during the fall and winter; it is used as a soil spray to stimulate root growth, humus formation, and microorganism development.

Preparation 501 is made from powdered quartz, also packed inside a cow horn and buried for six months, but during the spring and summer. It is applied as a foliar spray to encourage photosynthesis and plant growth.

Preparations 502 to 507 are herb-based compost inoculants, made from yarrow blossoms, chamomile blossoms, stinging nettle, oak bark, dandelion flowers, and valerian flowers. These five preparations, one made from each of the above plant materials, are used in making biodynamic compost, a key element in biodynamic soil fertility.

Preparation 508 is applied as a foliar compost tea, created from dried horsetail plant, and utilized for disease control (e.g., fungicide).

Again, the biodynamic ideal is that all of these preparations are made from materials grown on the farm. In practice, most of these preparations are purchased or made from purchased materials. In the United States, the Josephine Porter Institute for Applied Bio-Dynamics is the primary commercial supplier of these preparations to farmers.

8

GOING ORGANIC, OR
SOMETHING LIKE THAT

THE ROAD LESS TRAVELED

Conventional. Sustainable. Organic. Biodynamic. Beyond Organic. The spectrum of farming systems one might embrace includes these and others that lie between or maybe even outside their boundaries. As budding farmers, we had to find our own path among these many choices. I'll be the first to admit that in the beginning we hardly knew enough to know what the choices were, much less what our farming philosophy would be. During vineyard development and the first few years of operations, we delegated the decision making to a vineyard management company hired to manage our vineyard for us. More or less by default, we delegated the farming philosophy as well. There was no menu of choices presented anywhere in the process to choose organic, sustainable, biodynamic, or any other set of principles and practices. What we got as a result was somewhere between conventional and sustainable, with farming practices driven primarily by cost and the standard practices of the firm we had hired. In general, I think this served us fairly well initially, but our perspective had changed considerably by the time we moved to the vineyard and took up residence full-time.

Our vineyard was well into its fourth growing season, referred to as "fourth leaf" in the business, when we jumped with both feet into full-time farming at the end of May 2005. We had been preparing for this transition for several years. We were ready, or so we thought. What I know now, that I didn't know then, is just how fast vines can grow in May and early June. At their peak, vigorous grapevines can grow nearly two inches per day for two weeks or more and average over an inch of new growth per day for a month or so. Overnight, or so it seemed at the time, the vineyard went from neat little upright shoots about a foot tall to a jungle of shoots more than four feet tall. A thousand vines per acre, with sixteen to twenty shoots per vine, were growing wildly, inside and out of the trellis wires. It was completely and utterly out of control.

It was one thing to come up to the vineyard on weekends and tuck in a few hundred unruly shoots that escaped the shoot positioning work of the vineyard management crew; it was another thing all together for two novices to rein in more than fifty thousand out-of-control shoots on three and half acres of vines. After flailing for a week or so, losing ground in the process, we called in the professionals to bail us out. In just a couple of days, a small crew from our former vineyard management company had come in and managed to get all our vines back up and inside the trellis wires.

Yes, we were humbled, but by no means defeated. Out of that experience came an obsession with planning and tracking vineyard operations. There would be no more surprises (yeah, right!). To this day, every activity in the vineyard is recorded in detail. One spreadsheet holds a month-by-month overview of vineyard activities, including the specific tasks performed and the details of any materials that may have been used. Another spreadsheet details all irrigation inputs, such as the hours of irrigation and gallons of water applied per vine each week. A third spreadsheet provides a detailed hourly accounting of all our labor inputs for each week of the year. There are, of course, even more spreadsheets for various other details, but you get the picture. As the old saying goes, if you don't measure it, you can't improve it.

That first year of living and working in the vineyard also set the stage for a dramatic change in our farming philosophy. However, it may be more accurate to say that it led us to explicitly consider and implement our own philosophy, as we really did not have one beforehand. As we worked in the vineyard everyday, handling the vines, eating the grapes, or just sitting on the ground and petting our dog or playing with our cats who roamed and hunted in the vineyard, we became acutely aware of the potential impact of all the materials and practices we used in our vineyard. So it was that I began to question many of the precepts of our farming operations.

WEEDS AND COVER CROPS

One that stood out was the rather barren strip of soil lying under the vines in each row. Herbicides, primarily Roundup, were proving quite effective in eliminating any and all plant life underneath the vines. The idea behind this practice is to eliminate competition for water and nutrients. Where initially I found a certain beauty and symmetry in the alternating strips of bare earth and lush ground cover that adorned so many vineyards, I soon came to view that bare strip as a wasteland, devoid of the organic matter and microbial life so essential to healthy soil.

It would take a year or so before I fully came to grips with the consequences of herbicide usage in the vineyard, and it would be several more years before I fully grasped the consequences of eliminating herbicides. With only an inkling of where we were headed, after that first season of living and working in the vineyard, we decided to stop using herbicides. I knew at that point that a move toward organic farming was the right thing to do and that eliminating herbicides was the right first step.

I would learn much later that my transition to organic farming was to take a different path from most. For many, if not most, vineyards that transition to organics, the elimination of herbicides is the final and most difficult step. Rarely is it the first step. Of the four major categories of agricultural chemicals used in viticulture—herbicides, fungicides, insecticides, and fertilizers—herbicides are generally the most difficult to eliminate or replace with materials approved for organic production.

Managing noxious weeds and otherwise unwanted vegetation growing under the vines is a perennial, season-long problem for almost every vineyard. Herbicides, sprayed in a strip two or three feet wide directly under the vines, have proven to be a very cost-effective solution to this perennial problem. The alternatives are invariably more costly and time-consuming. If you don't attack them with chemicals, your options are basically to mow them down or dig them out. You can mow them down with machines, such as string trimmers, also known as weed whackers, or with grazing animals such as sheep. You can dig them out manually with a hoe or by machine with special-purpose tractor implements that basically plow them out. You don't need to be a rocket scientist to figure out that removing acre upon acre of weeds with weed whackers or hoes is a physically demanding and time-consuming task.

In contrast to all that manual labor, using sheep to mow down the weeds and grasses might not sound like much work. But sheep? Really? Finally, tractor-driven implements would seem like a good solution. I mean, how hard could it be to drive the tractor up and down the vineyard while it digs out and destroys the weeds? If only it were so easy. And did I mention how much these special machines cost? Or how much tractor fuel costs? Or how often the machines break? Or how many vines they chew up along with the weeds? Believe me when I tell you, weed control is the bane of every organic winegrower.

Nonetheless, we plunged into our second season of full-time grape growing without a Roundup safety net. Unfortunately, a very wet March and April made it impossible to get a mower into the vineyard until May, when the grasses in the rows between the vines had reached shoulder height in places. When I was finally able to mow in May, the enormous grasses were mown and deposited in a dense thatch that blanketed large sections of the vineyard. A few months later we discovered that this blanket of thatch made for vole (meadow mice) heaven, providing the shelter that enabled them to fearlessly burrow through the thatch to the base of our vines, where they munched on the tender tissues of the vine trunks just under the bark. That spring we lost at least a dozen vines to the voles.

Later that fall, faced with the perennial challenge of wet spring mowing and undervine weed control, a novel, totally organic solution caught my attention. Sheep. Olde English Babydoll Southdown sheep, to be exact. A few adventurous growers in the area had begun to experiment with Babydoll Southdown sheep to manage vineyard weeds and grasses during the winter and early spring prior to bud break. Short of stature and voracious grass eaters, the Babydoll Southdown sheep could maneuver freely and easily throughout the vineyard, going under the drip lines, fruiting wires, and other vineyard infrastructure.

Forgetting for a moment that we had no prior experience with livestock, the first challenge was finding sheep. Olde English Babydoll Southdown sheep are known as a heritage breed. These are the livestock equivalent of heirloom tomatoes, only much more rare. I figured I needed ten to twelve sheep to manage my four acres. I might buy ten or twelve sheep if I could find them, but their rarity and heritage breed status enabled breeders to charge four to five times the going price for the more common, but much larger, standard breeds of sheep. Trolling the Internet for "Babydoll Southdown Sheep for Sale," we happened upon a small flock of seven sheep for sale by a reluctant shepherd who could no longer afford to keep them. The price was good, and they were located only about a hundred miles away. Not the ten or twelve sheep we needed, but maybe a start. With three ewes and four rams, we figured that we could grow our flock to a dozen sheep in no time and would soon be selling lambs to recoup our investment. We bought them, sight unseen and without any of the infrastructure needed to manage, contain, or protect them. Buy sheep first, ask questions later.

One of the questions we did have the foresight to consider before we bought our sheep was, How are we going to protect them from predators? Mountain lions, coyotes, and even neighborhood dogs are mortal threats to sheep. Though we had never actually seen one, we had good reason to believe there were lions living in the mountains above and around our property, and we knew there were plenty of coyotes around, not to mention dogs of all shapes and sizes on the neighboring ranches and residential properties. While many small sheep ranches use llamas or donkeys to protect sheep and goats from marauding canines, we knew the mountain lion threat would require stiffer measures. We found the answer in a breed of livestock guardian dogs, called Maremmas, native to central Italy. Although rare in the United States, Maremmas have been used to guard sheep and goats from wolves, bears, and even thieves in the rugged mountains of Italy for centuries. Maremmas are large dogs with amazingly gentle temperaments and an instinctive ability to bond with and protect other animals with which they live. They learn quickly who belongs in their flock and who does not and will ferociously defend their flocks and territory from any intruders.

As soon as we committed to the idea of bringing sheep onto our ranch, before we even located any sheep to purchase, we went in search of a Maremma to be their guardian, whenever they did eventually arrive. We found a Maremma puppy only a couple hours' drive away, on a goat ranch in the Santa Cruz Mountains, recently born to a pair of Maremmas that had emigrated from Italy to California barely a year earlier. As a puppy, he would be more nuisance and troublemaker than guardian for the first year or so, but Francesco, named after the beloved Italian patron saint of animals, would eventually become an indispensable protector of our sheep and a dear friend as well. However, we decided that as a puppy he would be better suited to spending his first few months on the goat ranch where he was born, learning the guardian trade from his mother and father. So, Francesco was not yet living here when the sheep arrived, and for better or worse, he missed quite a bit of excitement.

All of a sudden we were winegrowers and sheep ranchers, and two of the three ewes we acquired with our small flock were presumed to be pregnant. Not only had I never birthed a lamb, but it had been twenty years since my cat delivered kittens, which was my most recent experience with live birth. As for Deb, well, she invoked that famous line from *Gone with the Wind:* "I don't know nothin' 'bout birthin' babies." What had we gotten ourselves into? We didn't know how much we didn't know, but at the time it just seemed like the right thing to do. Eliminate herbicides and reduce fossil fuel usage while mowing the vineyard naturally and fertilizing it with sheep urine and manure at the same time. How organic is that?

We parked our new flock of sheep temporarily in one of our neighbor's pastures, as they had several pastures through which they rotated a flock of several dozen goats. We checked on them a few times daily as we prepared our own property for moving the sheep into our vineyard. A couple of weeks later, on the day before Thanksgiving, we got a sobering call from our neighbor. A mountain lion had killed and carried off three of their goats in the pasture next to our sheep! Here we were, a couple of hours before dusk, faced with leaving our sheep totally exposed to a marauding mountain lion. We needed some kind of shelter. Fast! To make matters worse, whatever we came up with had to serve for at least two nights since no stores would be open on Thanksgiving Day to buy materials for building a new shelter. Frantically we rallied around our only solution. We had already purchased a few hundred feet of movable, plastic electric fencing that would eventually be part of the fencing used to contain the sheep in our vineyard.

Those first two nights we corralled the sheep inside several concentric circular layouts of battery-powered electric fence, lit the area with a big spotlight, and put out a portable radio blasting a talk radio station all night long.

The sheep survived, and we were probably more stressed than they were. I know we slept less. On Friday after Thanksgiving we went to our local home-improvement store and bought the materials to construct a makeshift eight-by-ten-foot barn. Thereafter, every evening we would herd the sheep into the barn for the night, returning each morning to let them out to graze. All this, and they had yet to even set hoof in our vineyard!

Nature has a way of testing rookie farmers, and our mountain lion ordeal was only the beginning of our education. Pretty soon, mowing and weed whacking was going to look pretty attractive! Shortly after solving the mountain lion dilemma, we came over one evening to put the sheep inside for the evening, only to find two of the rams locked in mortal combat. The foreheads of both sheep were battered raw and bloody from hours of head-butting battle. None of the ewes were in heat, so we don't know what set them off, and we had no idea how to stop it. For hours on end, the two rams, father and son, would stand head to head, periodically backing away ten steps or so and then crashing together in a violent display of the behavior that earned male sheep the ram designation.

Hours of research turned up the only solution apart from physically isolating the rams, which was not suitable given our need to contain all the sheep in a single pasture or vineyard block. The answer: ram shields, which are leather masks that block the forward vision of the rams. Turns out that if they cannot see straight ahead when facing their opponents, they get confused and will not back off and charge. So we ordered two masks to be shipped overnight, and then the fun started.

Now sheep are naturally skittish and not amenable to human contact or handling. So I grabbed my shepherd's crook and went about chasing the rams to fit them with their new masks. One ram went

Our sheep at work in the vineyard. (Photo by Deb Kiger.)

along fairly easily. The other wanted nothing to do with me or the mask. After a long chase and a diving catch, I managed to grab him around the neck. This big guy weighed over two hundred pounds and was a lot stronger than I was. I hung on for dear life and eventually wrestled him to a halt, though he dragged me around for five or ten minutes before giving in to the inevitable. With considerable difficulty, and very little cooperation, I finally got his mask on and breathed a sigh of relief. Sure enough, the two rams went right back to their head-to-head confrontation, but the masks worked as advertised, and the butting stopped. After watching their confusion for a few minutes, we walked away satisfied the problem was solved.

Unfortunately, one of the rams proved to be Houdini in wool and repeatedly escaped from his mask. Consequently, the aforementioned chase and wrestling match was repeated numerous times, each episode as comical as the one before it. Eventually I gave up, and we soldiered on with only one ram wearing his shield. That mostly kept the problem at bay, but we realized having four rams was not going to work and quickly put three of our rams up for sale. Sadly we were soon down to four sheep—one ram and three ewes. Fortunately, two of the ewes were pregnant, and we were hopeful that within a month or so we would have at least two and probably four little lambs running around to replace our unruly rams.

Even with the crazy mountain lion and head-butting episodes, the benefits of our ovine mowing crew were visible within weeks of moving them into the vineyard. In next to no time at all, they had mowed the cover crop to a finely manicured carpet of grass. Also, by the time we moved the sheep from the neighboring goat pastures to our vineyard, Francesco had come to live with us—or, more precisely, to live with the sheep in the vineyard. Within a few months, as Francesco matured and proved his mettle as a guardian, we were able to dispense

Francesco at work in the vineyard, guarding his sheep. (Photo by Deb Kiger.)

with the nightly enclosure and allow the sheep to graze around the clock.

As we were down to only four sheep, the grasses began to outgrow the grazing crew by late February. We knew help was on the way with two pregnant ewes, but we had no idea when the lambs would arrive. To make matters even more perplexing, these sheep are naturally so rotund and thick with wool by the end of winter that ewes are often not visibly pregnant until the final few weeks of their pregnancy.

After breakfast one Sunday morning in late February, I looked out the window into the vineyard to see Francesco visibly excited and circling around two small wriggling forms on the ground next to one of our ewes. Oh my God! We had twin lambs! And she did it all by herself!

Three weeks later, on another Sunday morning, our second pregnant ewe went into labor. We could see through binoculars that she was in labor and appeared to be expelling the water sac that generally precedes delivery. As we watched from a distance for an hour or so with no progress, we decided closer inspection was required. Approaching the ewe we found that the water sac we saw through the binoculars was actually the head of a lamb. A normal lamb birth occurs with the front feet coming out first, or the front feet and nose coming out together. In this case, the head was all the way out and the front legs were stuck inside the womb. Although this "two-headed sheep" was almost comical, the reality was a serious, life-threatening situation to both the ewe and the lamb.

As you might imagine, getting emergency veterinary assistance on a Sunday morning is not easy. After a tense fifteen or twenty minutes, we had a return call from the vet who attempted to talk us through an assisted delivery. Her instructions: coat one arm from fingers up to the elbow with a lubricant and reach in through the birth canal, locate the misplaced front legs, and carefully pull them out. Oh, and by the way, don't push the lamb's head back into the womb because it will probably suffocate. I don't know about you, but reaching through a ewe's vagina into her womb and fishing around for missing front legs sounded incredibly intimidating. Doing this with the full head of a lamb in the way just seemed impossible. So I delegated this task to Deb, reasoning, "Your hands are smaller. It will be a lot easier for you." Needless to say, she was as intimidated as I was. With the vet on the phone providing guidance, we gingerly fished around for the missing front legs, but ultimately the situation proved to be more than we could handle. We pleaded for the vet to come out and help us.

A long, tense hour later she arrived. I had spent the hour on my knees holding the ewe stationary as we attempted to calm her and prevent any accidental harm to the as-yet unborn lamb. In the meantime, Francesco had spent the hour licking the lamb's face, clearing away the birthing fluids and helping the lamb to start breathing on its own.

After a quick scrub and lube, the vet went to work. It took a while, but she found the legs, repositioned them, pulled them out under the lamb's nose, and then pulled the rest of the lamb all the way out. He lay there on the ground lifeless. We had saved the ewe, but the lamb appeared to be dead. She picked up the lamb by the back legs and swung him around in circles, attempting to expel any fluids that might be blocking his airways. We laid him back on the ground and watched as he magically came to life and began breathing on his own! In no time he was up on his feet and looking for mama and something to eat. We were all three nearly in tears. It was the tensest, most heart-wrenching drama I had ever experienced. Because of his oversized head and related birthing problems, we named him Big Head Todd the Monster, known affectionately today as Todd.

With Todd and the other two lambs we were back up to seven sheep and on our way to filling out our flock. But we soon learned that nature can be cruel, and despite their hardy nature, sheep can die in many ways. Over the next couple of years, we delivered several more lambs but managed to lose just as many sheep along the way. Our ram succumbed to a veterinarian's error administering a sedative for a minor surgical procedure. One of our ewes died from pregnancy complications, taking two unborn lambs with her. A ewe lamb, less than one year old, died of hypothermia after she slipped into a water-filled ditch and spent the night half-submerged. Another ewe had to be given away due to birthing complications that rendered future pregnancies life threatening. A second ram died of unconfirmed internal injuries, related as best as we could tell to an accident that occurred during mating. And finally, even with the protection of Francesco, two lambs succumbed to predators when they wandered off during the night away from the other sheep and out of the protective sight, smell, and hearing of Francesco.

Without a doubt, baby lambs are among the most adorable creatures on earth, and we found the whole experience of raising lambs incredibly exhilarating and rewarding. However, we also discovered that most of the mental stress, complications, and expenses associated with our ovine vineyard mowers were related to procreation. Furthermore, due to a combination of inexperience, accidents, and misfortune, after four years we still had only six sheep: five neutered males and one partially lame ewe.

So in 2010 we decided to discontinue our sheep breeding. If we need more sheep, we'll just buy them. We had tempered our goal of ten or twelve sheep, learning that we could manage our most difficult vineyard block with the six sheep we have. That leaves a small block to manage entirely with machines, but it also provides a comparison against which we have been able to judge the effectiveness of our sheep program. We have found that the sheep reduce our mowing and weed whacking by at least 50 percent, often much more. The benefits are greatest in drier years when there is relatively little cover crop growth after the sheep are removed from the vineyard in April. Nearly seven years into it, and having eliminated the complications of breeding and reproduction, we have found that maintaining a small flock of sheep is virtually no additional work; they require little attention or care and do a great job of keeping down the weeds and grasses in the vineyard.

FERTILIZERS

Around the same time that the Roundup-scorched bare strips of earth under my vines set me off on the sheep path, I also decided that the chemical fertilizers we had been applying were actually poisoning the soil and would eventually do more harm than good to the vines. Several factors conspired to produce this conclusion.

On the one hand, despite regular applications of high-potassium and high-phosphorus fertilizers, we were still experiencing chronic deficiencies of both nutrients in the annual analyses of leaf petiole tissues. At the time, we also speculated that these nutrient deficiencies contributed to the extensive leaf reddening we observed well before harvest each season, a problem we ultimately determined was caused by viruses rather than nutrient issues.

A second and more telling observation came from the soils analysis we conducted. A typical application of fertilizer in a vineyard is by injection of the fertilizer into the drip irrigation system, which delivers the fertilizer in a stream of water dripped under the vine. This is called fertigation. Typically a system will have one or two drip emitters per vine, generally placed a couple of feet to one or both sides of the vine trunk. Water and fertilizer are dripped slowly onto the ground, wetting and fertilizing a small patch of ground near the vine. The problem arises with the continued application of these fertilizers to the same spot over and over, year after year. As it turns out, only a small portion of the fertilizer dissolved in the water is readily available to the vine. In the case of phosphorus and potassium, most of these key nutrients become chemically bound up with other elements in the soil and are unavailable for uptake by the vine. Over time, the potassium and phosphorus salts that comprise the fertilizer build up in the soil. High levels of any salts in the soil are not healthy for plants or any of the microbial life a healthy soil depends on, and our soils analysis showed hot spots of potassium and phosphorus building under the vines.

Thus we faced a dilemma: how to provide the essential nutrients the vines needed without poisoning the soil in the process. The answer to this dilemma was found in a fundamental rethinking of our approach to vine nutrition. The conventional approach to fertilizer relies on feeding the vines concentrated nutrient solutions as directly as possible, generally via fertigation. The soil in which the vine is growing is almost an afterthought, as though it were an inert material existing only to support the vine and deliver the concentrated nutrient solution to the vine. The alternative we adopted was to focus on feeding the soil, rather than the vines, so that the soil could then feed the vines. I liken this distinction to the difference between relying on vitamin supplements for nutrition versus eating a healthy, balanced diet of natural foods.

Think about it this way. In our vineyard, vines are spaced five feet apart in each row, and the rows are nine feet apart. That gives each vine forty-five square feet in which to grow and search out water and nutrients. Over time, the vine will consume and deplete the nutrients in those forty-five square feet if there is no source of replenishment—thus, the siren song of chemical fertigation, as it seems such a simple and elegant solution to replenishing the nutrients in a concentrated spot of soil easily within reach of the vines' roots.

Even apart from the buildup of fertilizer salts under the drip emitters, there are other deficiencies in the chemical fertigation approach. For example, as in the vitamin supplement analogy, plant nutrients in chemical fertilizers are often not provided in forms easily taken up by the plants, and excess nutrients are often applied just to meet the plant's basic needs. Furthermore, it is very difficult for synthetic fertilizers to provide the full spectrum of nutrients that plants need, and as the plants deplete the broader soil profile, they become increasingly dependent on the small patch of soil fed by fertigation.

Our shift to feeding the soil that feeds the vines led us to focus on the entire forty-five square feet of soil available to each vine. Our goal was to develop a rich soil food web that would support the nutrient needs of the vines through decomposition of organic matter on or near the soil surface and by the cycling and redistribution of minerals and nutrients available in the air and soil. The soil food web consists of the biological community living in and on the soil, including the plants, bacteria, fungi, insects, and even our sheep. This soil food web is essential to the conversion of minerals and organic matter into the nutrients that the vines need to grow and produce a crop each season. By feeding and nurturing this soil food web across the entire floor of the vineyard, we would be ensuring that the vines had the nutrients they needed when and where they needed them.

Our first step in developing this soil food web was to promote the development of a lush carpet of beneficial grasses and low-growing broadleaf plants that would cover the entire vineyard floor. Between the rows of vines we seeded grasses and clovers to supplement the native grasses, and we also seeded the previously bare strips under the vines with low-growing perennial grasses. These grasses, clovers, and companion plants, such as the native filaree, provide a carbon rich cover crop that grows during the rainy season but dries up during the summer so as to minimize competition with the vines for water and nutrients. The cover crop also helps maintain a porous, aerobic soil structure and minimizes erosion that would leach away important vine nutrients. The residues from sheep grazing and mechanical mowing of this cover crop provide organic matter for the soil, and the sheep provide important recycling of the nutrients in the cover crop into forms that can be consumed by the vines and other elements of the soil food web.

The second key step we took was to provide an annual supplemental feeding with a layer of rich, aerobic compost broadcast across the vineyard floor. The compost provides supplemental organic matter, minerals, and nutrients as well as beneficial microorganisms that make up a healthy food web. In the first several years, we have also been amending the compost with mineral supplements such as rock phosphate, potash, and gypsum to overcome any deficiencies indicated by our biennial soil analysis and annual plant tissue analysis. Over time we hope to reduce or eliminate the amendments, relying solely on the compost nutrients and the ability of the soil food web to unleash the nutrients naturally occurring in the vineyard air and soil.

Converting a vineyard from chemical fertigation to an organic-soil-focused program of nutrition is not always easy, especially when starting with poor or depleted soils. In our case, with a young vineyard, depletion was not an issue. However, our rocky soils were very low in organic matter, which did present a bit of a challenge, and it has taken several years to build up the soil to naturally support the vine vigor and productivity we desire.

Which brings me back to the topic of fertigation. One need not forgo fertigation as part of an organic program; in fact, there are many scenarios in which fertigation needs to be an integral part of a vineyard nutrition program. Organic fertilizers derived from fish, kelp, and other organic sources can be important nutritional supplements, especially where the soil is incapable of meeting the needs of the vines. Fertigation with organic fertilizers can have some of the same drawbacks as chemical fertigation, especially when one is relying on fertigation for the bulk of vine nutrition. However, in drought years, when a large area of the vine's root system goes dormant early as soil moisture is depleted, nutrients delivered with irrigation can become very important. Similarly, when transitioning from depleted or deficient soils, supplemental vine nutrients beyond that which the soils can deliver become especially critical. We used liquid fish fertilizer for several years as we were building up the soil with compost, though we have been reducing our usage as the nutrient balance in our soil has been improving with our compost program.

Over the course of five or six years, we have seen a positive transformation in both our leaf petiole tissue samples and our soil samples. The petiole tissue samples, taken at bloom each year, show the vines to have an adequate and balanced nutrient profile, and we have eliminated the chronic potassium and phosphorus deficiencies that existed when we started this conversion. Likewise, our biennial soil samples have trended toward a more balanced nutrient profile and, most importantly, show a significant improvement in soil organic matter. Perhaps the most visible manifestation is the transformation of the vineyard floor during the winter rainy season, from a sparse collection of native grasses, noxious weeds, and bare soil into a lush green carpet of grasses and clovers.

FUNGICIDES AND POWDERY MILDEW

I have to say that the transition to organic mildew control was probably easier and more seamless than weaning off herbicides and chemical fertilizers. At the same time, it was more intimidating and fraught with peril. If your herbicide withdrawal goes south, and you start to freak out over an exploding weed population, salvation is just a jug of Roundup away. It would hurt to admit defeat, but the actual consequences would be relatively minor. Even letting the weeds run rampant probably would not be consequential, in the short term anyway. The fertilizer withdrawal could have collapsed in a similarly painless reversion to fertigation. Humiliating, maybe, but we'd probably be no worse for wear. Screwing up the mildew program, on the other hand, can carry significant, if not devastating, consequences.

Organic mildew-control programs require more frequent spraying and more thorough coverage of the foliage than conventional mildew controls. Canopy management—specifically, keeping the canopy open for sunlight and airflow—is especially critical for organic mildew programs. The fungicides used in organic programs are less powerful (i.e., less effective) and require direct contact with the mildew for effect. Conventional fungicides, on the other hand, are systemic. That is, they are absorbed by the vine tissues and transported through the vine to other tissues to confer protection even where direct contact of the spray material on the vine does not occur. This is a powerful mechanism, but it also means that the vine essentially ingests the fungicide, and much as drugs in your bloodstream are carried to all parts of your body, so are the fungicides in the vine . . . including into the grapes in many cases.

The organic contact fungicides are certainly sprayed onto the surface of the grapes, but the materials used, such as sulfur, mineral oil, and potassium bicarbonate, seem much less intimidating or dangerous than the synthetic chemicals making up the systemic fungicides. Also, most of these organic materials are photo- and biodegradable, breaking down naturally during the two months from the end of the mildew spray season to harvest and winemaking.

So it was a different perspective that motivated the move to organic fungicides, as compared to the herbicide and fertilizer decisions. With the herbicides and fertilizers, the primary focus is on the soil, how these materials affect the soil, and how the soil affects the vines. In the case of fungicides, the focus shifts to the grapes and potential residues in or on the grapes at the time of winemaking. Also, pesticide residues on the vines are a source of concern for those of us handling the vines on a daily basis. It was probably the residues on the vines that really set me to thinking about using organic fungicides.

I first encountered this issue when tracking the fungicides being applied to our vines by the vineyard management company in the days before I personally took over the fungicide spraying. Each fungicide has a product label that is required as part of its registration with the Environmental Protection Agency and other pesticide regulatory agencies, which includes the various safety precautions required for usage of that particular fungicide. In addition to all the warnings about the need for protective clothing, gloves, goggles, and respirators, I was alarmed by two other precautions that are required for all pesticides—the reentry interval (REI) and the preharvest interval (PHI). The REI is the minimum time that must pass before anyone can enter the vineyard after a pesticide application. The PHI is the minimum interval that must pass before harvesting the grapes after a pesticide application. Typically, for the conventional fungicides that were being used on our vineyard, the REI was either twelve or twenty-four hours, and the PHI was fourteen days. Basically, they were saying do not go near the vines for up to twenty-four hours, and do not consume any of the fruit for fourteen days. The fact that on the day after a fungicide application, Deb and I might spend five to six hours handling the vines with our bare hands was a bit disconcerting, to say the least.

It was not lost on me that most of the organic fungicides came with an REI of four hours and a PHI of zero days, meaning that you could (presumably) safely pick and eat the fruit four hours after application of the fungicide. As I came to understand this better, it strengthened my conviction that a complete move to organic farming was undoubtedly the right thing to do.

I would add that in all three cases—fertilizers, herbicides, and fungicides—there is a larger environmental perspective at work, driving the overall move to organic production. There are unquantifiable benefits from using softer materials and more natural controls like the sheep. There are also psychological benefits—it just seems like the right thing to do. However, there is a downside, captured neatly in Deb's characterization of "organic" as a code word for "more manual labor required."

The same season we moved away from chemical fertigation, we also switched to organic fungicides. Yes, we did have to spray more frequently, and the cost of our materials went up as well. Not all organic fungicides result in higher material costs, but there is no way around the more frequent spraying. That first season we were still contracting out the fungicide sprays, even as I was managing the program of what would be sprayed and when. Nonetheless, the dramatic increase in costs that first organic season prompted my decision to bring the fungicide program in-house. So at the end of that first season, I bought a tractor and an air-blast sprayer, figuring I could recoup the cost of the sprayer in only a few seasons. As for the tractor, well every farmer needs a tractor!

Fortunately, the transition to organic mildew control went off without a hitch. Several seasons went by without so much as a spot of mildew to be found anywhere. That would all change in 2010, but that's a story for a later chapter.

CONTROLLING OTHER PESTS BY VARIOUS MEANS

Weeds and mildew are not the only pests that bedevil vineyards, many of which I describe in chapter 5, "If It's Not One Thing, It's Another." In some cases, as with the European grapevine moth, the infestation and treatment is (hopefully) a onetime episode rather than a chronic condition. Fortunately, organic treatments for the European grapevine moth are available. They are much more time-consuming, requiring four to six sprays for organic growers versus two when using conventional, systemic insecticides. The European grapevine moth is the only insect pest for which we have ever engaged in any kind of control or eradication program.

No doubt we have various leaf-eating and grape-sucking insects. Many vineyards are plagued with mites, thrips, sharpshooters, and other little bugs that feed on grapevines and their leaves and fruit. In some cases these infestations are related to specific cultural practices in the vineyard. For example, using sulfur for powdery mildew control can exacerbate mite problems by eradicating beneficial mite predators. Similarly, soil tillage in the vineyard exposes the bare earth to sun and wind, creating dusty conditions that promote mite infestations. We do not use sulfur or tillage and have been spared any noticeable mite problems.

The insect populations that do exist in our vineyard are generally benign or beneficial, such as ladybugs. The few injurious insects we do encounter do not create problems that are large enough to warrant treatment with insecticides. The most serious insect problem we experience is with bees in the fall just before harvest, when they are attracted to the sugars in the ripe grapes. I'm never sure how much grape damage is from the bees feeding on the grapes alone or to what extent the bees are just moving in on the grapes that have been punctured by the various birds that feed on the grapes just before harvest. There is not much we can or are inclined to do about the bees. Honeybees, in particular, are having enough problems these days without us getting after them. Yellow jackets are also a problem, in which case I am a bit less sympathetic and opt for the basic hardware-store-variety yellow jacket traps.

Birds have been a problem for us and are nearing a damage threshold where some response is warranted. We have tried the easy solutions. The simplest and most common tactic is to tie shiny Mylar ribbons to the tops of the stakes in the vineyard. The ribbons blow in the wind and reflect sunlight, supposedly inhibiting the birds from alighting on the vines. I've tried it a couple of times, as it is cheap and easy. Frankly, I don't think it has much effect as the birds acclimate to the ribbons pretty fast. Another common approach I have been using is an electronic bird distress-call machine. It is a small box with a speaker and some electronics that generates a variety of bird distress calls, targeting bird species that are known to flock to vineyards for ripe grapes. There is some independent empirical evidence that these machines do reduce bird damage in some cases, but in no way do they eliminate the problem. My experience is that the birds acclimate to the sounds just as they do to the ribbons and that some species of grape-eating birds simply ignore the sounds altogether. Other growers use a similar sound machine that generates booming gunshot sounds rather than bird distress calls. I suspect that the results are similar to the bird distress-call machines.

The only really effective solution to bird problems is netting. As the grapes begin to ripen in the fall, netting is draped over the vines to keep the birds out. You can buy special equipment that spools out the netting and drapes it over the vines, but it still requires at least three people to operate the equipment and help lay out the nets. Needless to say, applying and subsequently removing the netting is very time and labor intensive—not to mention the cost of acquiring and maintaining the netting and associated equipment. Our bird damage has not risen to the netting threshold, but it is getting close. I can only hope the birds move on to someone else's vineyard before it comes to that.

Burrowing animals have been our most significant animal pest problem. For most vineyards that means gophers. For us it means ground squirrels and voles. When the ground was being prepared for our vineyard planting, we inadvertently created an ideal ground squirrel habitat. One of the factors that make our vineyard ideal for growing high-quality wine is the rocky soil, providing living testimony to the old adage that grapevines have to struggle to produce great wine. Our soil is more than 50 percent rock, from small pebbles to giant boulders five or six feet in diameter. Many of these rocks came to the surface when the ground was being prepared for planting and were pushed into rock piles that line the hillside along the down slope at the edge of the vineyards. Here the ground squirrels thrive in the protective scree of the rock piles, safe from most predators and only a quick scamper away from the vineyard. While they live in the rock piles, they prefer to bear and raise their young in underground burrows, which they love to dig in the vineyard. If left unchecked, the vine rows would look like Swiss cheese, chock full of squirrel holes. These squirrel holes and burrows present risk of leg injury to humans and sheep working in the vineyard, and they often collapse into sinkholes that make for rough tractor work during mowing and spraying.

I encountered the squirrel problem early on, shortly after we moved to the vineyard full-time. I first noticed the problem when I found a mound of dirt building up in a vine row not far from the house. The mound was probably a foot high and nearly three feet in diameter. The hole from which it was being excavated was more than eight inches in diameter, right at the base of a vine trunk. I was baffled and initially figured I had a badger or some similarly large creature living in my vineyard. The only clue to the contrary was that there appeared to be other smaller holes and mounds nearby as part of a network of burrows. I never saw anything coming or going, so I decided to camp out and see who the resident critter was. I pulled a chair out, sat twenty or thirty yards away, and waited. That was when I discovered I had a ground squirrel problem.

The easiest and most effective solution would be poison, using toxic baits to attract and kill the squirrels. Though legal, I find that option appalling. Granted, ground squirrels are classified as pests, and in California and much of the western United States, they harbor fleas infested with *Yersinia pestis*, the bacterium that causes bubonic plague. But the collateral damage of using poison is unacceptable, with potential by-kill of our neighboring predators and scavengers, including raptors, vultures, crows, skunks, raccoons, foxes, bobcats, and mountain lions, as well as our own dogs and cats.

So I am locked in a never-ending war with the ground squirrels, armed with lethal but relatively ineffective weapons, given the prolific fecundity of the native ground squirrel population. In addition to digging up my vineyard, they feed on the grapes as they are ripening in the fall. One ground squirrel can strip the grapes off an entire vine in an afternoon, and we often lose all the grapes on the vines at the ends of the rows facing the rock-scree paradise inhabited by the squirrels. Armed with a rifle and various traps, with tactical support from our resident hawks and owls, we do battle each summer and fall. Mostly it's a standoff. They get lots of grapes, and I get a few dozen squirrels.

Burrowing enemy number two is the vole, or meadow mouse. These little critters also live in networks of burrows dug under the vines. Because their burrows are small, there is relatively little structural damage to the vineyard floor. The problem with voles is their fondness for the tender vine tissue lying just under the bark. When a vole chews on the vine trunk, it typically works its way around the vine trunk near the ground. Unfortunately, the tender tissue the voles like to eat is also the vine tissue that carries water and nutrients from the roots up to the leaves, grapes, and other growing tissues. The vole damage cuts off the vine's nutrient and water supply, and the vine either dies or is unable to ripen its grapes.

Shooting or trapping voles is simply not practical, and as with squirrels, the use of poison baits is not acceptable. Our permanent cover crop, for all of its other benefits, likely contributes to the vole problem as it provides cover for the voles as they move from their burrows to the vines. This is where the sheep help tremendously. By keeping the grasses short in the winter, the aboveground habitat is unfavorable for vole mobility and also exposes those that do move about to predators. Vole damage has been dramatically reduced since we began grazing the sheep, though the number of burrows visible in the vineyard suggests a thriving vole population.

In response to the apparent population boom among the voles, we added two vineyard cats to our menagerie in 2011. The cats live with Francesco and the sheep in the vineyard. Within days of introducing the six-month-old kittens to the vineyard, we were witnessing vole captures by the little feline hunters. Often when we are out working in the vineyard, the cats will follow us up and down the vine rows, occasionally dashing off in pursuit of a vole spotted a few rows away.

One of our goals with the sheep and cats is to add a bit of biodiversity to our farm and leverage the natural behaviors of these animals to the benefit of vineyard. To that end we also have a few chickens that cohabitate with Francesco, the cats, and the sheep. The chickens spend their days roaming the vineyard, eating grasses, seeds, and bugs, recycling organic nutrients for the vineyard in their own way. The steady supply of wonderful free-range eggs is a great benefit for us as well.

Amazingly, all of our vineyard animals coexist peacefully and even provide mutually beneficial services. Francesco is the protector of them all, keeping the many predators at bay that might otherwise seek to make a meal of any one of them. We have also seen the chickens walking around on the sheep or standing beside them picking insects off the wool on their backs and around their faces. All in all they are quite entertaining and provide functional as well as aesthetic value to our vineyard.

WHERE'S THE ORGANIC VINEYARD SIGN?

By and large we have adopted a farming philosophy and set of practices that conform to the guidelines of organic farming as codified in the US Department of Agriculture's National Organic Program. However, we have chosen not to seek formal certification for organic production. There would be no economic benefit to us with certification. Our winery partner already pays us a premium price for our grapes, recognizing the value of our site and our farming practices, and the additional costs and paperwork associated with certification would just add to our overhead.

There are many who argue that the standards and regulations governing organic farming have been diluted in recent years to accommodate the needs and wishes of large-scale organic farms. The industrialization of organic farming has certainly brought compromises, not all of which I would agree with. Nonetheless, these standards have dramatically improved the agricultural landscape and in general promote more environmentally conscious farming practices. I like to think that we are taking advantage of the best practices and safer materials that the organic standards provide, with the freedom to go beyond those standards and practices where we can. Our use of animals for pest control and nutrient cycling is a case in point. So for us, organic farming just seems like the right thing to do and provides a lifestyle that we cherish and feel good about. Even if we don't have a sign or certificate to prove it.

9

THE ART AND SCIENCE
OF MILDEW CONTROL

Regardless of whether one chooses to grow wine using conventional, sustainable, organic, or biodynamic practices, a mildew-control program is a given. For the most part, you cannot wait until you find a mildew infection to treat or implement mildew controls. For all practical purposes, if you wait until you see mildew, you are in a heap of trouble.

There are four major facets to any mildew-control program: (1) managing the vines to minimize mildew risk, (2) monitoring for mildew infection, (3) selecting the fungicides to spray on vines to prevent and control mildew, and (4) deciding when and how often to spray these materials.

VINE MANAGEMENT

There are two important aspects of vine management that affect mildew risk: vine vigor and canopy density.

It is well documented that excess vine vigor contributes to numerous disease and pest problems in wine grapes. Insect pests, for instance, gravitate to high-vigor vines. Powdery mildew also thrives in high-vigor environments. Vigorous vines are characterized by heavy, dark-green foliage that is high in the nutrients on which insects and mildew thrive. The primary cause of overly vigorous vines is excessive fertility.

While planting vineyards in deep rich soils can be the cause, excess nitrogen fertilizer is more commonly the culprit. The old saw that the vines must struggle to produce great wine means that great wine is seldom produced in high-fertility soils. Nonetheless, some vineyards are planted in deep, rich soils, and managing vigor in such sites can be a problem. There are viticulture practices to mitigate the deleterious effects of excess soil fertility, but it is difficult to produce great wines under such conditions. These vines will also be more prone to disease problems. This is a problem that is best avoided during initial site selection, before planting the vineyard. Once a vineyard is planted, the blueprint for vine vigor is mostly determined by the site, but vigor can be controlled to a degree by managing fertilizer and irrigation inputs judiciously. So fertilizer and irrigation inputs become important determinants of wine quality and disease susceptibility.

The second and more common vine-management issue for controlling powdery mildew risk is canopy density. Even well-balanced, low-vigor vines produce enough leaves to create a well-shaded interior that is highly conducive to mildew growth. The best way to combat this is to maintain good light penetration and airflow into the canopy, especially in the fruiting zone. Ideally you want to be able to see dappled light when looking through the vine in the fruiting zone, which means there are scattered spots where you can see all the way through the vine. This is accomplished by removing leaves in the interior of the canopy so that light and air can penetrate. Many growers pull only the lateral shoots that grow in the fruiting zone, leaving all the leaves on the main shoot stem, allowing light, air, and fungicide spray penetration while also removing the most susceptible young leaves from the fruiting zone. This is particularly critical for organic and biodynamic growers since the organic fungicides must contact the mildew directly to be effective.

In particularly cool growing areas or mildew-prone sites, many growers will remove all of the leaves in the fruiting zone. This strategy is effective for minimizing mildew risk; however, sunburn risk then becomes a significant factor. In warm to hot climates, or where rows run in a north-south orientation, direct sun exposure creates significant sunburn risk. At a minimum, excessive sun exposure bleaches out color from the skins (and resulting wines), and in the worst cases it can cause premature raisining of exposed grapes and even complete loss of exposed clusters. A compromise employed by many growers with north-south row orientations, myself included, is to remove leaves to expose the fruit on the morning (east) side of the rows but not remove any exterior leaves on the west-facing side of the vines.

MONITORING FOR MILDEW

It is often said that the most effective pesticide is boots on the ground. This means walking through your vineyard looking for trouble. Like with most pest and disease problems, early detection is key to controlling and containing mildew outbreaks. In general, by the time you can see mildew infections, they have most likely already been established and growing for a couple of weeks. Nonetheless, without regular monitoring mildew infections that do occur will mostly likely spiral out of control by the time they are discovered.

FUNGICIDES FOR MILDEW CONTROL

While many would like to believe that organic crops are grown without pesticides, this is unfortunately often not the case, especially where wine grapes are concerned. Powdery mildew is a threat just about everywhere *Vitis vinifera* grapes are grown for wine, which means some type of fungicidal treatment will be required. In regions where rainfall is common during the growing season, downy mildew, black rot, and botrytis are also problematic. Choosing which fungicides to use brings you back around to the farming philosophy you subscribe to. Most commonly the fungicide choices are grouped into two categories, conventional and organic.

Fungicides, conventional and organic, are regulated in the United States by the Environmental Protection Agency (EPA) and state-specific agencies such as the California Department of Pesticide Regulation (DPR). In order to use any material as a fungicide on any food product, including wine grapes, the material must be registered as a fungicide with the EPA. To apply any such material in California, the material must also be registered with the California DPR. To be registered with the California DPR means that a specific product has been approved for usage against specific fungal diseases on specific plants or crops. The registration also specifies how much product is to be applied, minimums and maximums, how often it should be applied, and safety regulations regarding its usage and disposal. All applications of pesticides within the state of California must also be reported monthly to the California DPR. Similar agencies and regulations govern pesticide usage in other states within the United States, the European Union, and all other major winegrowing countries, such as Australia and New Zealand.

Conventional Fungicides

Conventional fungicides represent a vast and ever-changing array of synthetic chemicals developed specifically to combat one or more fungal diseases. Almost all of these are patented chemicals developed and sold by large agrichemical companies such as Bayer Crop-Science, Syngenta, BASF, Dow Agrosciences, and Monsanto. They are the backbone of mildew control in conventional farming but are also widely used in sustainable farming. None of the sustainable farming certification programs for wine grapes prohibit the use of conventional fungicides.

Conventional fungicides are grouped into categories based on their mode of action. The mode of action is the mechanism by which the fungicide affects the mildew organism, such as by disrupting cell division or disrupting nucleic acid (DNA/RNA) synthesis. This topic will be addressed in greater length later when we look at fungicide resistance in the chapter "Nature Strikes Back." Most of these are systemic fungicides, meaning that they are absorbed by the plant and translocated to other tissues.

Organic Fungicides

Organic fungicides are not, strictly speaking, organic materials in that they are not necessarily carbon-based compounds. Nor are they necessarily derived from living organisms, as the more common definition of organic might imply. Rather, terms such as "organic fungicide" and "organic fertilizer" refer to materials or products approved for organic production by regulating entities and programs such as the US Department of Agriculture's National Organic Program.

Furthermore, whether we like to think of them this way or not, most organic fungicides are generally considered chemicals. For example, the most widely used organic fungicide in vineyards is sulfur. Sulfur, of course, is a naturally occurring chemical element and indeed appears in the periodic table of chemical elements. Copper, another member of the periodic table, is also used as an organic fungicide.

What sets organic fungicides apart from the conventional alternatives are the types of chemicals involved and their general toxicity (or lack thereof). Organic fungicides are derived primarily from naturally occurring compounds, although they may actually be manufactured rather than mined or produced directly from naturally occurring sources. Here are the most commonly used materials in organic fungicides for controlling powdery mildew on wine grapes:

- Bacteria
- Copper (the Bordeaux Mixture [a solution of copper sulfate, lime, and water] as well as other fixed copper compounds such as copper hydroxide)
- Hydrogen peroxide
- Plant- and mineral-based horticultural oils
- Plant extracts
- Potassium bicarbonate
- Sulfur (sulfur in dry formulation [sulfur dust] and sulfur-water solutions)

DECIDING WHEN, HOW MUCH, AND HOW OFTEN TO SPRAY

There are basically two approaches that winegrowers use to manage their spray programs for controlling powdery mildew. The most common approach is to simply follow the instructions on the labels of the products being used. The label will indicate an interval between sprays (e.g., seven to fourteen days) and a quantity to be sprayed (e.g., two to four quarts per acre). The general implication is that the longer interval and lower quantity is to be used when mildew pressure or risk is low and the shorter interval and higher quantity when the risk or pressure is high, but the assessment of the interval and quantity to be used is left up to the grower. Grower surveys indicate that about half the winegrowers in California make this determination without any formal measurement of the mildew pressure or risk. Most such growers simply pick an interval and a quantity from the specified ranges and follow a regular schedule based on those decisions (e.g., two quarts per acre every fourteen days). This simple calendar-based approach to spraying is a conservative approach to mildew control but is generally not the most cost-effective or ecologically sensitive approach.

An alternative to the simple calendar-based spray program is to use an analytical or empirical model of mildew risk or pressure. The most widely used model is the Gubler-Thomas Powdery Mildew Risk Index, developed at the University of California at Davis in the mid-1990s. The model uses leaf wetness (e.g., rainfall, fog) and temperature to model the risk of initial infections and temperature to model the risk of the established infections that are the source for most mildew epidemics. The conidial infection index is the most widely used aspect of the model, particularly in relatively warm, arid winegrowing regions such as California and Australia. The index ranges from 0 to 100 in increments of 10, with values of 0 to 30 indicating low mildew risk, 40 to 50 medium risk, and 60 to 100 high risk. This index can be used in conjunction with the fungicide interval and quantity ranges provided on product labels to manage the grower's spraying program. For example, when the index reads 60 or higher, the interval between sprays should be at the minimum indicated on the label and the quantity of fungicide used at the maximum. In addition, the model has a trigger value, and before this trigger is reached, the mildew risk is presumed to be near zero. Using the trigger value as the signal to start the spray program for the season and subsequently managing spray intervals and quantities using the index reading enables the grower to make fewer sprays and to use less fungicide material than when using a basic calendar-driven approach.

Gubler-Thomas Powdery Mildew Risk Index readings are available on a daily or weekly basis to growers in California and other regions from various agricultural weather services. The index is also implemented in weather stations that growers can buy and deploy in their own vineyards, which provide more accurate, site-specific readings. I have one such weather station installed in my vineyard. Grower surveys in California show that about 30 percent of growers rely on the Gubler-Thomas model for all or most of their mildew spraying intervals.

The Gubler-Thomas model has been found very useful in California and other winegrowing regions of the world with similar warm, arid climates. However, it has been shown to be less effective in some other climates, notably very cool climates such as Tasmania and climates that receive considerable rainfall during the growing season. A similar model, called OiDiag (e.g., Oidium Diagnostic), has been developed in Germany and is in widespread use in European vineyards. The OiDiag model uses daily temperature data but also factors in humidity, leaf wetness, rainfall, and ontogenic resistance data to calculate an index for determining the interval between sprays.

Whether a spray program is based on the calendar or one of the mildew forecasting models, it has become clear in recent years that there is a critical period of susceptibility to mildew infection, especially where the fruit is concerned. This period runs from one week before flowering until about thirty days after flowering, when the berries are two millimeters in diameter, and has been called the open window period. Research has shown that preventing infection during the open window period is the most critical factor in minimizing, if not eliminating, infected fruit at the time of harvest. Using conventional fungicides, it has been shown that three well-timed sprays during this period are as effective as a full season of seven sprays at controlling mildew on the fruit at veraison (and hence harvest). It is unlikely that organic growers could get away with only three sprays during the open window period, but using shorter intervals and higher spray rates during this critical period would seem prudent. The implication is that control of mildew during this critical period will keep your fruit clean all the way to harvest, and any subsequent sprays can be reduced based on observed mildew pressure and desire to keep foliage clean to minimize overwintering mildew populations.

IO

NATURE STRIKES BACK

f or every action there is an equal and opposite reaction. While Newton's third law of motion might not apply to biology and evolution, there does seem to be a corollary that makes much of nature so unpredictable. For any action there may be an unintended reaction. Indeed, this is a rather apt description for many of mankind's attempts to harness or redirect nature. A rather poignant example can be found in the history of antibiotics. Prior to 1935, internal infections such as pneumonia or those resulting from injuries such as surgical wounds and other types of puncture wounds were generally untreatable and often fatal. With the discovery of sulfa (sulfonamide) drugs in the early 1930s and the subsequent invention in the early 1940s of a means to produce penicillin in useful quantities, such infections were no longer life threatening. Both discoveries led to Nobel Prizes and fueled an antibiotic revolution that dramatically altered the practice of medicine and human welfare in general.

Even in the earliest days of the antibiotic revolution, there were warnings about the unintended consequences of wanton antibiotic usage. Penicillin was introduced to widespread medical usage in 1943, and by 1946 there were reports of penicillin-resistant infections. As more and different antibiotics were introduced, the pattern continued to repeat itself. After a period of unrestrained efficacy of the antibiotic, drug-resistant strains of bacteria would appear. Some antibiotics had longer runs than others, but inevitably resistant strains developed. Streptomycin was used for fourteen years before resistant infections were reported, while methicillin resistance was reported the year following the drug's introduction.

Drug-resistant bacteria, the unintended reaction to these antibiotics, are the result of evolutionary mechanisms that have evolved over millions of years. The discovery of penicillin came about from the accidental observation of the effects of certain mold cultures on bacterial samples in a laboratory. The antibiotics streptomycin and tetracycline were both derived from compounds produced by streptomyces bacteria. Not all antibiotics have been produced from natural sources such as these, but the mechanisms by which resistance develops share much in common. Unfortunately, the same mechanisms that drive resistance to antibiotics interfere with our attempts to control powdery mildew using the synthetic chemicals known as systemic fungicides.

Way back in 1875, the eminent Scottish farmer and poet Robert Burns penned these immortal words: "The best laid schemes of mice and men go often askew." Yes, indeed.

FUNGICIDE RESISTANCE

Until the introduction of systemic fungicides in the 1960s, the control of grape powdery mildew had been provided by sulfur- and copper-based fungicides, dating back to the 1850s and the original oidium epidemic in France. Although effective, these fungicides require frequent spraying and direct contact with the mildew. As a result, they require a lot of labor and tractor fuel, and effective application becomes increasingly difficult as the vine canopies grow larger during the season. Those limitations were offset somewhat by the fact that there do not appear to be any mechanisms by which the fungus (*Erisyphe necator*) develops resistance to the antifungal properties of the sulfur- and copper-based sprays.

The introduction of systemic insecticides in the 1950s set the stage for systemic fungicides, although it would be a couple more decades before any such products came to market. However, even with the early systemic insecticides in the 1950s, signs of resistance to the pesticide properties emerged early on, as amply documented in Rachel Carson's landmark book *Silent Spring* in 1962.

DuPont introduced the first systemic fungicide for powdery mildew control in 1968. DuPont's Benomyl provided both curative and preventative treatments for mildew infections. Benomyl enabled growers to extend the intervals between spraying and freed them from their worries over complete or perfect spray coverage. These attractive benefits led some growers to abandon sulfur- and copper-based sprays completely, relying solely on Benomyl for protection. However, in as few as three years from its introduction, reports began to surface of resistant strains of mildew.

Benomyl targeted a specific protein involved in cellular mitosis, a key step in the cell division process by which the genetic information is transferred from one cell to another, a fundamental building block of development and growth for animals, plants, and fungi. Random mutations in the gene responsible for the targeted protein reduced the sensitivity of the mildew fungus to Benomyl's target mode of action. In organisms such as *Erisyphe necator*, which can undergo many reproductive cycles in a relatively short time frame, such mutations can rapidly lead to large populations of resistant fungi, which is precisely what happened. Furthermore, as turned out to be the case with Benomyl, these mutations can persist in fungal populations for ten years or more without any subsequent Benomyl exposure. Hence, the resistance is very durable once established.

Nearly all systemic fungicides target a single biological function, typically at the molecular or genetic level, which is referred to as the mode of action. Numerous other fungicides using the same mode of action were introduced subsequent to Benomyl. Though they may have differed somewhat in the specific biochemistry employed in each fungicide, their common mode of action meant that resistance developed to any one of them was transferable to any other.

The Benomyl experience has proven to be the rule rather than the exception for systemic fungicides. The introduction of the demethylation inhibitor (DMI) fungicides in the vineyards of California in the early 1980s provides a case in point. The first DMI fungicide was registered for use on grapes in California in 1982. Most growers were still using sulfur to control mildew at the time but were not happy with its limitations (i.e., frequent applications, phytotoxicity if applied during heat waves, negative affects of sulfur residues on wine quality). As a result, many growers abandoned sulfur for the new DMI fungicides, and reports of DMI-resistant mildew began to surface as early as 1985 and were confirmed in research reports a few years later.

Fungicide resistance is not an insurmountable problem for the wine industry, but it does depend on the diligence of both farmers using the fungicides and the industry that creates them to insure that the resistance risks are understood and that the fungicides are not misused in practice.

Today all fungicides are categorized according to their mode of action and the target site that the fungicide acts on. An industry-sponsored organization, the Fungicide Resistance Action Committee (FRAC), manages the categorization of fungicides. In the Benomyl case, the mode of action is mitosis and the target site is the b-tubuline protein involved in mitosis. Appropriately, as Benomyl was the first systemic fungicide and the first to uncover the resistance issue, the FRAC code for Benomyl and related fungicides is 1. There are more than sixty FRAC codes in use today to describe the resistance properties of fungicides.

Where fungicides are categorized as having a medium or high risk for the development of resistance, which is true for all of the systemic fungicides used against grape powdery mildew, it is imperative that growers follow usage protocols designed to minimize the risk of resistance developing in mildew populations. The product labels are required to describe these protocols as part of the instructions for using the fungicide. Typically the protocols specify not only the allowable amounts of fungicide to be used in a spray mixture (e.g., ounces per acre) but also how often the fungicide can be used in a single season, what other fungicides it may be mixed with, and other restrictions that may exist on how it should be used in a rotation of several different fungicides during the course of the season. The primary weapon used to fight the development of resistance is to rotate the use of different categories of fungicides, so that resistant fungi that survive the spraying of a particular category of fungicide one week will succumb to the use of a different category of fungicide two or three weeks later.

Fortunately, fungicide resistance in powdery mildew is not nearly so dangerous as the antibiotic resistance occurring in various strains of bacteria that are injurious to human health and life. Unlike with antibiotic resistance in bacteria, we have other tools for managing fungicide-resistant mildew in vineyards. The old standards of sulfur- and copper-based sprays remain as effective as ever. In addition, none of the other fungicides used in organic grape production show indications of susceptibility to resistance or are affected by any resistance developed in response to the use of systemic fungicides.

II

IT'S ALL-OUT WAR

A SEASON UNDER SIEGE

The dark side of mildew control truly eluded me in my first few years as a full-time farmer. That all changed in 2010. I might even go so far as to say that prior to 2010, I had been deceived, though it would probably be more accurate to say I was just naïve. I will say deceived, though, because everything that I had experienced up to that point had led me to believe that I really knew what I was doing. In hindsight I now understand better what happened and how preceding events led to my demise. Nonetheless, 2010 caught novices and grizzled veterans by surprise, and by no means was I alone or even in the minority.

In the spring of 2010, while no longer a novice I was certainly less of an expert than I wanted to believe at the time. In fact, my track record looked pretty darn good. I had four full seasons of running my own mildew-control program under my belt and had experienced only one small infection on a cluster of four Cabernet Sauvignon vines in 2009.

The 2010 season began in fairly ordinary fashion. Bud break came in late March, which I would say is typical for our Syrah, amid what had been a mostly warm and sunny month. There was a little rain early in the month and again at the end, but altogether it was perfect weather for bud break.

My first mildew spray typically occurs in mid-April, when the vines have new growth of three to six fully expanded leaves. Though it's usually too cool for the mildew to grow or reproduce at that time of year, that first spray knocks down any overwintering colonies that might appear and new infections waiting to spring to life from ascospores released during spring rains. This early spray also protects against colonization of primordial buds that are forming on the new shoots, which would lead to flag shoots in next year's early growth.

With ample rainfall after bud break, the April spraying seemed like a particularly good idea. However, the cool and damp conditions had put a damper on vine growth, and it was the end of April before we had three to six expanded leaves on most vines. By this point, we were moaning a bit about the lingering cool, damp weather but remained confident that the typical warm, sunny weather of spring and summer could not be far away.

With the new shoots only three to ten inches tall, spraying with the full tractor spray rig would have been overkill, and the wet ground would have made that problematic as well. So, as I usually do with the first spray, I used a small spray tank set in the back of our trusty six-wheeled John Deere Gator and drove up and down the rows, spraying the front and back side of each row with a handheld spray wand—time-consuming, with two passes for each row, but pretty effective.

All was going well until I carelessly backed the Gator off the edge of a terraced vine row and got stuck on top of a vine in the third row from the top of the vineyard. Couldn't go forward and couldn't go backward. I was stuck!

After noodling over the problem for a short while, I called my neighbor, who brought over his Gator and a winch. But we didn't have anything to secure his Gator to, so that didn't work. The winch just dragged his Gator toward mine when we tried to pull it out. So

I got out my tractor, which is much bigger and heavier than the Gator. We hooked the tractor to the back of my Gator, and his Gator/winch to the front, and managed to pull the Gator out without breaking any vines. What an adventure! So, a small spray job that should have been done in a couple of hours stretched out to take most of the day. It took longer to free the Gator than the whole spray operation otherwise.

The next morning I sprayed the lower vineyard Syrah with the same setup, minus getting stuck. This type of work needs to be done early in the morning, before the winds pick up. Otherwise, even light breezes can put as much spray on the operator as on the vines.

The Grenache was very slow to awaken in its second season. Most shoots were no more than a few inches long on May 6, well behind the Syrah. However, enough leaves and flower clusters were out to warrant the first spray. Given how little growth there was to spray, I used a backpack sprayer and knocked it off in less than an hour, using the same 1.5 percent Stylet Oil spray that I had used a week earlier on the Syrah. Stylet Oil is a clear mineral oil that is an excellent eradicant for powdery mildew and especially effective early in the season when the canopy is small and achieving full coverage is no problem.

Once all the vines had their bud break spray, I turned my attention to the Powdery Mildew Risk Index. I track the index using a software module in the weather station in my vineyard. As is customary, the index was set to zero at the beginning of the season. The index remains at zero until certain trigger conditions occur—specifically, three consecutive days each with six or more consecutive hours with temperatures between seventy and eighty-five degrees (and not to exceed eighty-five degrees either). The implication of the trigger is that until the trigger criteria are met, conditions are not favorable for mildew growth or reproduction. In theory, one need not even spray until the trigger is reached, which was my standard practice. Some years, it can be four weeks or more between the bud break spray and the first spray after the index is triggered. However, in the 2010 season the index trigger was reached on May 4, just before the first Grenache spray, so game on as of May 4.

May brought more rain and continuation of the cool temperatures that had so delayed early vine growth and the first mildew spray. Although there were four rainy days in the two weeks following the May 4 Grenache spray, cool temperatures prevailed and kept the mildew index down in the low-risk readings. As I monitored the index daily, my standard practice was to lengthen spray intervals when the index said the risk was low or medium and to shorten the intervals when the index said the risk was high. This program had worked nearly flawlessly for all of the four years I had employed it, and so 2010 started with more of the same. Either bravely or foolishly, I stretched my interval out to twenty-one days for the Syrah and seventeen for the Grenache, with twenty-one days being my maximum interval for Stylet Oil when the mildew risk indications were low.

By this time, all the growers in Northern California were moaning and groaning about the weather. It was early June, and we were two to three weeks behind schedule. Many varieties had not even begun to bloom. Our Syrah, which typically blooms in mid-May, did not show any signs of bloom until June 7. Fears of high disease pressure and a late harvest began to seep in. The weather, which had been unseasonably cool and wet, had kept the mildew at bay, but the incessant rains meant there was lots of mildew out there just waiting for some warm weather to kick into high gear.

With all the rain and cool weather, vine growth was running several weeks behind schedule, but by early June the vines were full tilt in hypergrowth mode. Flush with rainwater and sunshine, the vines were growing nearly 1.5 inches per day by the first week of June. The new growth was more than three feet tall, and the canopy was quickly becoming a jungle. A perfect storm was brewing, fed by the unusually rainy May and June weather and the dark, humid jungle of a lush canopy.

With the cool weather and low index readings, I held out for seventeen days between the second and third sprays, which was probably pushing the safe upper limit under those conditions, especially given the three rainy days since the last spray. By then, the vines were quite large, so I brought out the tractor and spray rig, a fifty-gallon air-blast sprayer with five spray nozzles on each side. I used only two nozzles on each side, which covers a three-foot-tall vine canopy pretty well, spraying about forty gallons per acre of a 1.5 percent solution of Stylet Oil in water. Stylet Oil is very popular among organic growers because of its low cost, ease of use, and rapid microbial breakdown in the vineyard with little or no residue left on the grapes or vines.

In recent years I had trended down toward 1 and 1.25 percent Stylet Oil solutions, with an eye toward using as little fungicide material as I could get away with, but had decided to go back up to 1.5 percent after a little mildew snuck through on a handful of vines the prior season. In hindsight, 1 percent versus 1.5 percent probably didn't matter. I am now fairly certain that forty gallons per acre did not provide adequate coverage given the large canopy and massive buildup of powdery mildew inoculum from all the rains. At the time it seemed OK. Forty gallons per acre and extended spray intervals with low index readings had served me well at bloom time in prior years.

But I should have known business as usual was not going to work in 2010. Nothing about 2010 was normal. I even predicted the coming problem but didn't realize it or act on it at the time. Here is an excerpt from my notes on May 29, 2010: "I think the cool spring has finally broken. It's 80 degrees today, and warm weather is forecast as far as we can see. That means all the powdery mildew inoculum born of the recent rains will explode in a flurry of reproductive activity by the end of next week."

If I made such a statement today, my subsequent behavior would be quite different from what it was in 2010. Live and learn.

With the wild canopy growth, we were removing leaves and lateral shoots constantly, trying to keep the canopy open. The vines were bigger, leafier, and greener than we had ever seen at bloom time, even though it was all happening much later than normal. In our limited winegrowing experience, we had yet to encounter a season as tardy as 2010. In 2006, the last pre-drought year in California, we had a very wet spring that delayed bud break and bloom. I witnessed the first bloom in 2006 on June 4, which seemed late at the time and proved to be several weeks later than what we observed in subsequent seasons. June 4, 2010, had arrived with no sign of bloom on the Syrah, Cabernet, or Grenache. The scariest part was the end-of-season implication. In 2006 our harvest occurred on October 28, compared to our more typical (and desirable) harvest dates around October 10, plus or minus a couple days. With each passing week in October, the threat of significant rain increases and with it the threat of any number of wine-quality issues. Still, the season was young, with plenty of time to make up for lost sunshine. In 2006, we made up some time with a brutally hot June and July, only to give it back with an unusually cool and cloudy September. We were saved by two weeks of sunshine and temperatures in the seventies and eighties in late October.

In 2010, after an unusual June 4 rain, our typical warm, sunny June weather moved in. Temperatures were mostly in the eighties and even wandered into the nineties. Good weather for grapes. Good weather for mildew too. As is typical for June in Sonoma, the mildew risk index was pretty much pegged on high. But I was sticking to my basic strategy of Stylet Oil and fourteen-day spray intervals. I noted all was well on June 28, with my journal reading, "So far, no sign of mildew."

I also noted at the time, "It's been fairly quiet in the vineyard." There was some pulling of laterals, moving of wires, and tucking of shoots into the wires before each spray. There was also some additional weeding and mowing, which was unusual for late June but part of the price we were paying for all the spring rains.

Mildew-free from 2004 to 2008, in late July 2009 I had found mildew had colonized all the grape clusters on four Cabernet vines in the top row of our vineyard. By the time I found it, the damage was done, and so I ended up dropping all the fruit from those vines onto the ground. Up to that point, I thought I was good, when I was probably just lucky. Maybe a bit of both, but all my studying, researching, and scheming to control mildew with a minimum of inputs had surely led to a bit of overconfidence. After all, as the old saying goes, if you think your vineyard doesn't have any mildew, then you are just not looking hard enough. It's always there, at some level, but hopefully under control and out of sight. Fortunately, the damage in 2009 was minimal, if not negligible.

So it was with some trepidation that we entered 2010, with the anticipation of setting our first Grenache crop. A couple years back, when we were researching the prospect of converting some Syrah to Grenache, a disturbingly common refrain was the description of Grenache as a "mildew magnet." My experience thus far had been only with Syrah and Cabernet, both of which are relatively resistant to mildew infection. It is surprising just how wide a spectrum of mildew susceptibility exists among different wine grape varietals.

It was Thursday morning, July 1, when I found it. I had been going through the Grenache vines for several days, pulling leaves and laterals to open up the canopy in advance of my plans to spray on Saturday. Everything looked good, and I had made my way through almost three-quarters of the Grenache when I found it: clusters of little green grapes covered in a grayish-white powdery coating. I started looking around nearby vines and found scattered infection on several more vines. The further away I looked from that first infected vine, the less I found. The infection appeared new and fairly localized. But immediate action was called for. Even though I planned to spray in two days and could have moved the spray up a day, I wanted to take stronger corrective action by washing the infected vines with an eradicant ASAP.

For organic growers, the most common eradicants used to cleanup mildew infections are Stylet Oil, which I use as my main fungicide, and Kaligreen. Kaligreen is an agricultural formulation of sodium bicarbonate. So the prescription is to take sodium bicarbonate to relieve the heartburn caused by finding the mildew infection, then to spray the vines with sodium bicarbonate (Kaligreen) to kill the mildew.

Turns out the sodium bicarbonate disrupts the potassium balance in the fungus, causing the cell walls to collapse. It's quite effective at killing mildew infections, but since it works only on contact, it does not provide any residual protection.

So Thursday afternoon I was off to our local agricultural supply house to get a bag of Kaligreen. Then it was off to the tractor dealer to get a handgun sprayer with a long hose that would allow me to walk up and down and between rows in the infected area and spray a heavy wash of Kaligreen solution directly onto the infected vines and grape clusters. The next morning I mixed up twenty gallons of water with six and a half ounces of Kaligreen and gave Deb her first tractor-driving lesson, and off to battle we went. Normally, spraying is a job for one person, who drives the tractor and operates the sprayer from the driver's seat. Spraying by hand requires two, one person driving the tractor and one walking behind with the spray wand. Everything powered by the tractor's PTO, as the sprayer is, requires a butt in the tractor seat. A reasonable safety precaution, I suppose. It took us most of the day to complete the job.

I followed up the hand spray with my regularly scheduled Stylet Oil spray the next morning. My thoughts at the time, as noted in my journal on July 3: "It's clear I will need to change my protocol for managing mildew in the Grenache. The 'same as Syrah' plan clearly doesn't work." Little did I know at that moment just how little the same-as-Syrah plan was going to work for anything, including the Syrah.

Three days later, on July 6, I discovered a mildew explosion in the Cabernet Sauvignon. Back out with the tractor and handgun.

While our first response to the mildew outbreak was to hand-spray the hot spots with a Kaligreen solution, the bigger task was to go through the entire vineyard and open up the canopy to air, light, and sprays—vine by vine, pulling leaves and laterals (small branches of multiple leaves that grow off the main shoots). Initially we focused on the Grenache and Cabernet, which were hardest hit, but ultimately we had to leaf all the Syrah as well. Given the right conditions, mildew will infect Syrah, tough as it is, and there was plenty to go around with all the mildew already growing in the Grenache and Cabernet.

We had been lulled into a sense of leafing complacency by the drought years. In 2007 through 2009, virtually no leafing was required. Our canopies were much smaller, due to low soil moisture during the spring growth period and some nutritional issues as we struggled to build up the soil in the early years of our organic transition.

We scrambled like mad to get the entire vineyard leafed in just a few long, hard days to get ready for the next spray. It took the two of us about forty hours of pulling leaves and laterals out of the fruit zone to get ready for the next spray—quite time-consuming, as it also entailed separating and repositioning misguided, tangled shoots and grape clusters. We cleared all the leaves on the east-facing side of the vines and thinned the west-facing leaves while trying not to fully expose the fruit to potential sunburn.

After discovering the initial outbreak at the beginning of July, I shortened my spray interval to ten days for the July 13 spray, in addition to hand-spraying the hot spots a couple of times in between. I also upped my spray rate to sixty-three gallons per acre—the most I had ever used. I mixed Stylet Oil and a bacterial biofungicide called Sonata to get their combination of eradicant and protective properties. Three days later mildew broke out in the lower vineyard Syrah and had spilled over into the upper vineyard Syrah as well. Events were spinning out of control.

So I shortened the spray interval to seven days, spraying the entire upper and lower vineyard again on July 20. I also continued hand-spraying the hot spots with Kaligreen in the interim. The next day, July 21, I noted in my journal, "Fortunately, the situation now seems under control."

Was I ever wrong! It was only a few days later, pulling laterals to get ready for the next spray, when I found a mildew wildfire raging in the lower vineyard Syrah. Of course, nothing was actually burning.

But I can think of no other more apt description of the situation. The mildew infection had exploded, from what had been a small hot spot a week earlier into a full-blown emergency encompassing nearly the entire block. I was devastated, and frankly bewildered.

I sat down right then and there, in the middle of the vineyard. What now? Nothing I was doing seemed to have any effect. Day by day the situation was only getting worse. Right there, sitting in the vineyard, I called the resident pest-control advisor (PCA) at the agricultural supply house where I purchased my fungicide materials. The PCA is their pesticide products technical expert and also has the benefit of working with many different growers and sees the big picture across many vineyards in the area. I explained my situation and pleaded for guidance. It was a small comfort to hear I was not alone and that organic growers all across Napa and Sonoma were being especially hard hit.

The prescription was daunting. My first order of business had to be stopping the mildew wildfire in its tracks. I needed to douse the mildew with a very high-volume eradicant spray. Specifically, two to three hundred gallons per acre of a combination Kaligreen and Stylet Oil solution applied right to the fruiting zone using my handheld spray gun. As the situation in the Grenache block was almost as bad as the lower vineyard Syrah, this was going to entail hand-spraying almost two acres of grapes. Since the tank sprayer on my tractor has only a fifty-gallon tank, I was going to need four to six full tanks per acre. Deb drove the tractor, no small matter given the treacherous terrain and her minimal experience as an operator, while I walked behind with the spray gun blasting the mildew. It was a very long, stressful day. It took four hours to get through the lower vineyard Syrah and another three hours to get through the Grenache.

Despite the breadth of infection, I was optimistic that we had arrested the problem before serious damage occurred to the fruit. But veraison was nowhere in sight. As the whole season was still running two to three weeks behind normal, it looked like veraison could be as much as three weeks away, especially for the Grenache. The battle would not be over until veraison was well under way. Confident, or at least hopeful, that I had knocked down the worst of it, I was still faced with preventing a resurgence for up to three more weeks. This brought me to the hardest, most painful decision I have faced in the vineyard, even to this day. The choice was put to me like this. I could continue my organic program, but scaled up to two or three hundred gallons per acre, once a week, to veraison, or I could bring in a conventional, systemic fungicide and probably make it to veraison with a single additional spray at about one hundred gallons per acre.

At that point, I had lost all confidence in my ability to control the current situation using the organic fungicides that I had been using. I also knew that if the mildew wildfire flared back up, I stood a good chance of losing much or most of the crop. After a couple of sleepless nights, I threw in the towel and decided to use a popular conventional fungicide called Mettle. I tank-mixed Mettle with the Stylet Oil and Kaligreen eradicants that I had been using for what I hoped would be a final blow to the mildew for the season, taking me into veraison with a reasonably clean and minimally damaged crop.

It turned out that veraison took so long to arrive that I did have to do one more spray in the lower vineyard Syrah and the Grenache in mid-August, but the situation was well enough under control for me to comfortably revert back to my organic materials. So, by the third week of August, a good three weeks late, we had good color in the Syrah, Cabernet, and Grenache. And the mildew had been defeated. We breathed a big sigh of relief. Home free, at last!

The mildew had been subdued. It looked like we would sustain some fruit damage but nothing like it could have been. We thought, let's wait a few weeks to get through veraison, and then we'll go out to drop any mildew damaged fruit. I expected we might drop a few hundred pounds, but at that point I was just ecstatic to have avoided catastrophe.

But Mother Nature was not through with us yet! Monday, August 23, dawned like most of that spring and summer, cool and foggy.

Temperatures started out in the forties that morning, but by midafternoon the thermometer had reached one hundred degrees. A bit shocking, but no harm done. Where was this hot weather in June and July, when we really needed it?

The next day was another story all together. Temperatures dropped into the fifties overnight and then exploded as the day warmed up. By mid-day it was over 100 and kept climbing to peak at 108 that afternoon. It was blistering hot all day long, with four to five hours that afternoon at more than 105.

This was looking eerily similar to July 2006, when we had a couple of consecutive days top out at 107. But in 2006 it had been hot all summer, and while we suffered some heat damage in the vineyard, the grapes had been toughened by the persistent hot weather, and overall the consequences were mostly cosmetic. It created extra work, cleaning up the damaged fruit, but in the end 2006 was our largest crop ever, and the wines from 2006 turned out beautifully. But in 2010, we had barely seen ninety degrees all summer long, and nothing had remotely approached what you might call a heat wave. On top of that, because of the cool, foggy weather and the resulting mildew problems, we had done more extensive leaf removal than ever.

It was still one hundred degrees at 7 p.m. when I ventured out to investigate patches of brown in the vines that I could see from inside the house. I figured some of the leaves had been scorched by the blazing heat. By the time I got within ten feet of the vines, I could smell the problem. Fruit had literally been stewed on the vines. Grapes were turning brown, cooked in place.

I should not have been surprised. A few hours earlier, with the weather station reading 108 degrees, I had ventured outside with my infrared thermometer. The deck chairs on my patio, facing west in full sun, registered over 130 degrees. The west-facing wall of our house—180 degrees! The same west-facing grapes never stood a chance.

A tour of the vineyard that evening and then again the next morning was depressing. It looked like 20 to 30 percent of the fruit was lost, literally cooked in the sun. A third consecutive day of scorching sun and triple digit temperatures didn't help. At the time, I could only hope for warm weather to follow so that the damaged fruit would dry up, rock hard, and not present any more problems (i.e., mold or rot). Fortunately, we had mostly sunny weather in the wake of the heat wave, and none of the damaged fruit turned moldy or rotten.

Over the next two weeks, the full extent of the damage played out in vineyards all across Napa and Sonoma. The immediate impact of cooked fruit was just the opening act. Within a couple of weeks, many vineyards, ours included, found widespread damage to the rachises of the grape clusters. The rachis is the stem structure that delivers water and nutrients to the grapes. As they shriveled and died, entire clusters of grapes were left to dry up and die on the vine. Clusters that had appeared unscathed in the immediate aftermath of the heat wave had succumbed to the rachis damage. What initially appeared to be a 20 to 30 percent loss of fruit now looked like it could be as much as a 50 percent loss.

It was time for damage control—to cut off the dead and damaged fruit. With a dramatically reduced crop load, we hoped that the vines would be able to make up some lost time and allow us to bring in the harvest sooner. Could that be the silver lining in all this devastation?

Most of September was spent cleaning up the mess from the August scorching. After completing the upper vineyard and maybe half the lower vineyard, I was happy to find that earlier fears of 50 percent damage were overly pessimistic. While damage in the lower vineyard approached 50 percent, the upper vineyard came out looking much better, with most of the Syrah in the upper vineyard in good shape. Finally, some good news!

After months of our bemoaning all manner of evil brought on by unseasonably cool weather from April to September, October came to the rescue. Unseasonably warm weather in the first two weeks of October saved the vintage. We decided to pick the upper vineyard Syrah on October 15 and let the remaining Syrah and Grenache hang for another couple of weeks.

By Monday, October 18, the weather forecast was looking rather ominous, with half our Syrah and all the Grenache still hanging. It looked like rain would start Thursday evening and continue almost straight through to Monday morning. With the specter of significant rain hanging over us, we decided to get the Syrah and Grenache in by Thursday, or Friday at the latest. The only problem was that practically every vineyard in Sonoma and Napa had exactly the same idea. Harvest crews everywhere were working ten or more hours a day to get as much fruit in as possible before the rain.

Larger vineyards that utilize these crews throughout the growing season for all or most of their farming needs naturally took precedence over our small, independent operation. The crew that had picked for us the prior week was booked to the max. I started calling around, contacting every vineyard manager I knew. By Wednesday morning, I had lots of "Sorry, too busy" replies and one maybe. So I started calling other vineyard owners to see if the people that picked for them could help. Still, by midday Wednesday I didn't have a crew to pick the next morning, and rain looked quite certain for Thursday night.

About 2 p.m. Wednesday afternoon, the maybe turned into a yes! A crew would be picking nearby in Kenwood starting at midnight and should be finished by 8 a.m. As we were looking at only about three tons, our job was one they could finish off in a couple hours after the Kenwood pick. A crew of about fifteen guys showed up around 8:30 a.m., and by 10 a.m. had picked nearly three tons of Syrah and Grenache. It was an amazing job by guys who had already been working almost nine hours when they got to our vineyard that morning. Shortly after lunch the grapes were safely on the ground at the winery.

Whew! What a relief. Bring on the rain! And rain it did, dropping five inches in just two days. But for us, a long difficult season was over. All around the story was the same: no one had ever seen a growing season like this one, and we were all very glad to get it over with.

20/20 HINDSIGHT

I can look back now on the 2010 season as an invaluable learning experience. All things being equal, I might choose not to have learned some of those lessons. But no one ever said farming would be easy, and if it were, it would probably not be half as much fun.

With the benefit of hindsight, I can now see the errors and omissions that contributed to our mildew nightmare. However, these insights did not come easily. I distinctly recall talking about the 2010 season with other growers early the next spring. At the time I was still somewhat mystified. Granted, the cool, foggy summer and thick, luxuriant canopies were widely recognized as the main drivers of the region-wide mildew problems. Myopically, I was somewhat stumped because of the poor correlation between the severity of the mildew season and the mildew risk index readings.

Hardened by my experience in 2010, I was a bit more prepared for a bad mildew season in 2011. The same basic weather pattern that brought us the mildew nightmare in 2010 was repeated in 2011, minus the encore heat wave of August 2010. I did not fully understand what was needed at the time, but I did shorten my spray intervals and increase my volumes in 2011 relative to 2010. That helped a lot but did not spare us all the mildew trials that 2011 had in store. As with 2010, organic growers generally had a tough time in 2011. Our Grenache, grafted over from Syrah in 2009, was in its first season with a full canopy and gave me a few mildew headaches, and the lower vineyard Syrah had a few problem areas as well. I caught them all early and avoided the panic and crisis of 2010.

It was only after a tough mildew season in 2011 and a new look at the weather and key events from 2010 and 2011 that the pieces fell into place for me. I already had an overlay of my spray dates on the Powdery Mildew Risk Index readings for the season as part of my basic record keeping. To this data I added (post–bud break) rainfall dates, the dates on which bloom began for each of those seasons and the dates when I first detected mildew. Since you have reached this point in the book, you know that rainfall produces infections, and the period of maximum susceptibility to infection is the period just before and after bloom. What 2010 and 2011 had in common were multiple, significant rainfall events in the week to ten days immediately preceding bloom. Add rainfall just before bloom to persistent cool, foggy weather with big, thick canopies, and you have the recipe for a major mildew headache.

12

FROM VINE TO GLASS

Growing and making wine unfolds each year as an intricate dance between man and nature. This is true for most of agriculture but is especially true for wine. This dance is more intricate for some wines than others. Wines grown and produced on an industrial scale tend to involve more human intervention, or manipulation as some might call it, than artisan wines. At the other extreme, the "natural wine" movement prides itself on a minimalist approach, striving for as little intervention as possible in farming and winemaking. Nonetheless, the specter of mildew infection hangs over almost every vineyard, requiring some intervention on the part of the grower to ensure that mildew does not ruin the grapes or the resulting wine.

Does this intervention in the battle to control powdery mildew affect the wines we drink? Are there perceptible effects to consumers and wine enthusiasts? Are there other implications of fungicide programs we should be concerned about?

There should be little doubt that the battle to control powdery mildew exerts a powerful and positive influence on the wines we drink. That is to say, without the tools and programs we use to control powdery mildew in our vineyards, there would be little or no wine as we know it today. This was amply illustrated more than 150 years ago with the first oidium epidemic in France and remains a fact of life almost everywhere wine is grown. However, this should not be taken as license to employ any and all means in the name of mildew control. As winegrowers (and winemakers) we have a responsibility to employ means and methods that respect our environment and the health of wine consumers.

The ideal for all winegrowers is to produce a mildew-free crop every year. However, the ideal is difficult to achieve, and it is reasonable to ask whether a little mildew here or there really makes a difference. In the worst case, the economic implications of unchecked or out-of-control mildew infections are fairly clear, with the loss of most or all of the grapes not out of the question. But what about the less-than-catastrophic mildew infection? If a little mildew is bad, is more even worse? Is there a threshold for mildew infection, below which everything is OK but beyond which only bad things happen?

Most winemakers will tell you they do not want any powdery mildew going into their wines. In reality, this is probably more of an aspiration than an actuality. But it's a worthy aspiration nonetheless. Mildew infections in vineyards can run a wide gamut, from the visually imperceptible to the fuzzy, moldy mess of rotting fruit. Severe, early-season infections can totally destroy a crop. Left unchecked, the mold will consume the fruit, resulting in a shriveled, rotting mess. Even if arrested fairly quickly, early-season infections can cause the fruit to split open as the berries expand, leading to desiccation of the berries and open wounds for spoilage yeasts and bacteria to colonize. Such fruit looks, smells, and tastes bad and is generally left to rot in the vineyard. That such fruit would be bad for wine quality is self-evident, if not just plain common sense.

Moderate to severe infections that occur later in the season can also be problematic in a number of ways. Fruit that is visibly mildew infected should be left in the vineyard, but that does not always happen. It may be because only small sections of some clusters are damaged, and it is judged too wasteful to discard all of the affected clusters. This puts the burden on the winery to decide what to keep or discard as grapes are sorted prior to crush. In other cases it may be that the severity of the infection is very light, in which case it may not be obvious in the field how extensive an infection might be.

When assessing mildew infections for quantitative purposes, they are typically described in terms of incidence and severity. Incidence describes or quantifies the percentage of grape clusters that are infected with mildew, regardless of severity of infection. Severity describes or quantifies the percentage of berries that are infected in those clusters. In cases where the incidence is high but the severity is low, a significant amount of mildew can slip past the harvest crew and even the sorting crew at the winery. Generally, when the severity is high, the mildew will be easily visible, even when the incidence is low. In large-scale winemaking facilities, any mildew that makes it from the vineyard to the winery probably goes into the fermentation tank. Most wineries producing premium, high-quality wines employ sorting operations that enable them to remove visibly mildew-infected or otherwise damaged fruit before it goes into the tank.

When mildewed fruit does get into the fermentation tank, there are several ways in which it can affect wine quality. On the one hand, it has been shown that mildew-infected fruit tends to have lower anthocyanins, a key determinant of color in red wines. Mildew-infected fruit has also been shown to differ in various measures related to wine aromas and mouth feel, such as the composition of phenolics and total acidity in the wine.

Mildewed grapes also tend to have a much higher overall microbial population than clean fruit, which can include yeasts and bacteria that influence fermentation, especially in the early phases before the primary fermentation yeast, *Saccharomyces cerevisae*, takes over. These non-saccharomyces microbes can introduce unpleasant sensory qualities to the wine, such as the fingernail polish aroma of ethyl acetate produced by acetic acid bacteria (Acetobacter) and the earthy, barnyard aromas of Brettanomyces yeast.

Sensory analyses and comparisons of wines made from mildew-infected fruit indicate that unpleasant or negative sensory qualities may be associated with powdery mildew infection and increase with increasing levels of infection. Effects observed in mildew-influenced wines include mold, fungal, or mushroom aromas and a pronounced oiliness or viscosity in the mouth feel. However, there does not appear to be a clear threshold for how much mildew infection is required for these negative sensory qualities to emerge. Whereas some studies have shown that negative impacts can be detected when as little as 1 to 5 percent of the grapes are infected, others have shown that up to 50 percent of the grapes can be infected with out perceptible flaws in the resulting wines. More often than not, these studies indicate that the negative effects of mildew on wine are perceptible at much lower levels of infection than 50 percent, typically when as little 10 percent or less of the fruit is infected. Unfortunately, most of these studies have not distinguished between the incidence and severity of the infections that went into the evaluated wines, but the overall impression is that negative sensory qualities can occur with even relatively minor mildew infections. Whether these negative sensory qualities are due to the powdery mildew itself or result from other spoilage organisms that often accompany mildewed fruit is not clear.

There is another, more insidious route for mildew to enter crush and fermentation during the winemaking process. It occurs when mildew infections are established late in the development of ontogenic resistance. Researchers at Cornell University have shown that in the late stages of mildew susceptibility, at about three weeks after fruit set, mildew infections can occur, but the ontogenic resistance has developed to the point where those infections are unable to grow and reproduce. Called diffuse powdery mildew, these infections are not visible to the naked eye but can actually cover significant portions of the cluster (berry) surfaces with a diffuse network of nonsporulating hyphae and microscopic necrotic blemishes. While these are qualitatively quite different from visible mildew infections, there is evidence from research conducted at Cornell University to suggest that even these diffuse powdery mildew infections can negatively impact wine quality. Grapes with extensive diffuse powdery mildew infections tend to have higher populations of spoilage microbes, higher levels of insect damage (which also leads to more spoilage microbes), and an increased tendency for bunch rot symptoms at harvest. While wines made from grapes with diffuse powdery mildew were not distinguishable on typical sensory attributes such as berry flavors, acidity, or mouth feel, they did elicit significantly more negative descriptors than the comparison wines made from clean fruit.

Another potential avenue for powdery mildew to affect wine quality lies in the materials used to prevent and control powdery mildew in the vineyard. Given that powdery mildew is controlled in vineyards using fungicides, it is not unreasonable to ask if these fungicides might have negative repercussions for fermentation, dependent as it is on yeasts, which are close relatives of the fungus that causes powdery mildew. Indeed, one might expect that fungicides applied to control fungi in the vineyard would have similarly deleterious effects on fungi in the winery during fermentation if present in sufficient quantities.

The key factor determining the impact of fungicides on fermentation would be the level of fungicide residues lingering on the grapes when they are brought into the winery. Where powdery mildew is the target of fungicides, residual fungicides are generally not an issue for fermentation. Fungicide applications for powdery mildew typically cease at veraison or earlier, typically fifty to sixty days or more before harvest. During this period, a wide variety of factors contribute to break down the fungicides. This is most notably true in the case of organic fungicides, which are generally not absorbed by the plant and remain on the surface for various environmental forces to act upon. Sunlight and heat cause many of the fungicides to break down or volatilize. Enzymatic or microbial processes degrade others. In many areas, summer rains wash the residues off berry surfaces, preventing or reducing accumulation.

In a few cases, particularly with the use of copper-based fungicides, the active ingredient (copper) does not break down or degrade over time. In this case, the residues will be a function of the amount of copper used in the field during the season and whether intervening factors such as rainfall have washed off any existing residues. Copper has been a part of many mildew-control programs since the Bordeaux Mixture was invented in France during the nineteenth-century downy mildew epidemic, and obviously a great deal of wonderful wine has been made since then from vineyards using the Bordeaux Mixture and other similar copper-based fungicides. When used in powdery mildew control programs for wine grapes, copper residues on the grapes are very low and have no effects on fermentation. However, studies have shown that if copper fungicides are used to control late-season disease problems such as downy mildew or various bunch rots, sprays applied close to harvest can severely impact fermentation and wine quality.

The most common concerns for fungicide effects on wine quality are associated with the use of sulfur. Dissolved in water or sprayed as a powder, sulfur has been used as a fungicide to battle powdery mildew since the original European oidium epidemic in the 1850s. Sulfur remains the most widely used fungicide for grape powdery mildew control and is the mainstay of many organic, biodynamic, and sustainable winegrowing programs. However, it is also well known that yeast in fermenting wine convert sulfur into hydrogen sulfide (H_2S), a nasty compound that smells like rotten eggs. Hydrogen sulfide may appear in wines during fermentation for many reasons and not necessarily as a result of sulfur residue from field applications of sulfur fungicides. Excessive or, more likely, late-season sulfur applications can leave sufficient residues to contribute to hydrogen sulfide production during fermentation. However, research has shown that sulfur residues on grapes in the vineyard decline significantly over the period from veraison to harvest and are unlikely to cause excessive hydrogen sulfide production or other sulfur-related wine faults. Nonetheless, many wineries ask growers to halt the use of sulfur, especially micronized sulfur, after fruit set for fear of lingering sulfur residues.

Finally, there is the subject of fungicide residue in wine itself. Pesticide residues are endemic in today's modern food chain.[1] This is true wherever conventional agricultural practices are employed and has become a driving force behind the growth of organic food production and consumption. A great deal of research has shown that nearly all conventionally grown produce, grapes included, contain some pesticide residues. Almost without exception, organically grown food products have been found to contain fewer pesticide residues than conventionally grown alternatives. Generally, the residues are well below thresholds judged to be safe by governmental regulatory agencies and do not necessarily pose a threat to human health or safety. While we rely on government and industry regulation to insure that pesticide usage results in safe or harmless levels of pesticides in our food chain, the fact remains that little is known about the long-term implications of low-level pesticide residues in our air, food, and water.

As with other fruits and vegetables, grapes are generally harvested containing some pesticide residues on the berry surfaces or even inside the skin, pulp, and juice when systemic fungicides are used. Where powdery mildew is concerned, various biological and environmental factors intervene to degrade the fungicide residues between the end of the mildew spray season and harvest. But some residues persist and do enter the winemaking process.

In arid winegrowing regions like California, few fungicides or other pesticides are used after veraison in most years. In other winegrowing regions, particularly where high humidity and rainfall are common between veraison and harvest, fungicides are commonly used even as harvest approaches. However, all fungicides are regulated with a

---

[1]The Pesticide Action Network has consolidated pesticide residue information from the US Department of Agriculture's Pesticide Data Program for many common agricultural products and related food products. This information can be found at http://www.whatsonmyfood.org

specific preharvest interval (PHI) that specifies the minimum period that must separate application of the fungicide and harvest of the crop. The PHI is meant to take into account the toxicity of the product and the expected rate of degradation of the material once it is applied. Typically, organic fungicides have very short PHIs and may even be applied on the day of harvest in some cases (which says a lot about their relative toxicity). Systemic fungicides, on the other hand, tend to have minimum PHIs of fourteen days or more—and up to sixty days in some cases.

All fungicides used in wine grape production are deemed safe by the agrichemical industry and governmental regulatory agencies when used in accordance with the product instructions and regulations. In the case of systemic fungicides, it is somewhat comforting to consider that the fungus-specific single-site mode of action by which these fungicides act on their fungal targets helps insure that they are not harmful to other organisms in the vineyard or to us when they linger in the wine and grapes we consume. Nonetheless, the unknowns associated with these chemicals lead many of us to prefer the generally safer and more benign effects of organic alternatives.

While there is relatively little research on the topic, it is clear that some pesticide residues do persist in some finished wine products. One study of forty European wines from France, Germany, and Austria published by the Pesticide Action Network of Europe found that all of the thirty-four wines they tested that were produced from conventionally grown grapes contained trace amounts of one or more pesticides, most commonly systemic fungicides. Where present, the pesticide residues were found in quantities ranging from a few to a few hundred micrograms per liter of wine, which is equivalent to a few or a few hundred parts per billion. In all but four instances, the residues were far below maximum safety thresholds specified by European Union regulatory agencies. While only six of the forty tested wines were labeled as produced from organically grown grapes, five of these six wines did not test positive for any pesticide residues.

13

THE ODYSSEY IN A BOTTLE

I am fond of the old adage that says wine is made in the vineyard. That's probably not much of a surprise coming from a grower, but most winemakers will support that sentiment as well. Make no mistake, the winemaker is a very important part of the equation, but even the most talented winemaker cannot craft a great wine from pedestrian fruit. Similarly, the grower also plays an important role, but ultimately it is the site and the particulars of the vineyard that set the bar for any given wine. The microclimate, soil, and sun exposure, along with the types of grapes being grown, are the factors that set the boundaries for any wine produced from a particular vineyard. So, however much I revel in facing the challenges we endure with each vintage, I constantly marvel at how fortunate we are to have stumbled upon such a great site for growing red wine.

The ultimate reward for all we do here in the vineyard is the wine. It has always been all about the wine. Granted, we embarked on our vineyard odyssey in search of a certain lifestyle, but without a passion for wine in the first place, the idea of owning and working a vineyard would never have happened. Among the most baffling experiences I have encountered as a grower has been meeting other growers who do not even drink wine. Heresy, I say. No surprise, teetotaling winegrowers are quite rare. Less rare, but no less baffling to me, are the growers who do not make, or have never made, any of their own wine. Actually, it's quite common to find growers who do not make any wine. Growing wine, they might say, is hard enough, and they are happy to leave the winemaking to others. But it seems to me that they are missing a crucial link in the circle of vineyard life, the link that closes the loop between the vine and the bottle.

My intention, from the day we hatched the idea that became our vineyard, was to make wine from the grapes grown in our vineyard. Of course, I knew no more about winemaking than I did about winegrowing when we started out, but this seemed somewhat less important as I had no plans to make wine for commercial sale. It occurred to me in the spring of 2002, after we had just planted a vineyard, and while I was still scaling the learning curve to become a successful winegrower, that I still knew nothing about winemaking. I realized that I was only a few short years from having my own grapes to work with, so I better figure out how to make wine. True, I could have waited to delve into winemaking once our vineyard was producing on its own, but I really wanted to be ready when I had grapes of my own. I did not want to waste my hard labor on a rookie winemaker!

In the beginning, it was winemaking by the numbers, and I had my nose in a book every step of the way. That first wine, in 2002, from purchased Syrah grapes, was not very good. Like any proud new parent, I did not see it that way at first, but over time I came to realize it was really not very good. Eventually, after dancing around the truth for a couple of years, we poured what remained of it down the drain and recycled the bottles for a new vintage. In 2003 and 2004 we bought Cabernet Sauvignon from a small vineyard near the town of Sonoma. We still have some bottles of each vintage, but only a few. I don't care much for the 2003, though Deb likes it OK. We haven't tasted a bottle in a couple of years, so who knows? The 2004, much to my surprise, was quite good when I opened a bottle earlier this year.

Each of these first three vintages was made in tiny quantities, less than ten cases each, with purchased fruit—as practice for the real thing, which was to come in 2005 with the first real harvest from Kiger Family Vineyard.

In 2005, with over ten tons of fruit harvested from our vineyard, we finally had the wherewithal to approach winemaking a bit more seriously. We kept almost half a ton of fruit for ourselves, crushing and co-fermenting a blend of 63 percent Syrah and 37 percent Cabernet Sauvignon for our first vintage from Kiger Family Vineyard. To shore up my winemaking credentials, I volunteered as a "cellar rat" in a local winery during the 2005 crush season. This allowed me to see firsthand how the pros made wine and gave me confidence that my own winemaking was on the right path.

That first wine from our vineyard turned out very well, and we are still enjoying it on special occasions to this day. We bottled about twenty-two cases of wine from the 2005 vintage, which brought us to the challenge of designing a label for our wine. This was not wine intended for commercial sale but rather for sharing with friends and family. But it still needed a name and a label. While working on a simple graphic design for the label, I began to brainstorm a name for our wine. Our vision was that every year the wine would be a bit different, mixing varying amounts of the Syrah and Cabernet Sauvignon grapes we grew in our vineyard to create a unique vintage each year. Simply calling it by the varietal names of the grapes that went into the wine seemed to lack proper gravitas. I wanted to capture something more meaningful, somehow more expressive of what went into the wine than simply the varieties of grapes included. That's when it came to me. We would call our wine The Odyssey to capture that sense of journey and adventure that lies behind each vintage.

Here, in the fall of 2012, we have just crushed the grapes for our eighth vintage of The Odyssey. I still find making wine to be an important link in the annual cycle of life in our vineyard. It connects us directly to the raison d'être for our vineyard, as wine is the reason we plant and nurture these vines as we do. It also provides a shared experience and vocabulary that connects us on another level with our partners and customers at Robert Biale Vineyards, especially Steve Hall, the winemaker at Biale. Most importantly, we are able to relive the ups and downs of each vintage as we drink The Odyssey over time, savoring the complexities that evolve in the bottle and remembering the quirks of nature that make each vintage unique.

The wine we make also connects us directly to our land and the way we farm our vineyard. I drink wine nearly every day, and most of it is wine we grow and make ourselves. This helps to put our winegrowing practices into perspective, in the knowledge that whatever pesticides or other materials we put into our vineyard will potentially end up on our dinner table in the wines we drink every day. This connection between our vineyard and the dinner table is a major factor in our commitment to low-input, organic winegrowing practices.

Over the years my winemaking has evolved in ways not unlike the changes in our winegrowing. It's not necessarily organic winemaking but definitely trends toward low-input and minimal intervention. In the early days of making wine by the book, I tended to make lots of adjustments and additions during the winemaking process. I sometimes added water to lower the potential alcohol in the finished wine or tartaric acid to lower the pH. I used various yeast nutrients to build a fat, healthy yeast population to ensure fermentations ran their full course, insurance against the headaches of stuck fermentations. None of these interventions are particularly controversial, outside the hardcore "natural wine" crowd, and they are in fact quite common in commercial winemaking. However, I never bought into some of the more esoteric wine manipulations used in many wineries today, such as enzymes for color extraction or added tannins to enhance certain sensory characteristics of the wine.

None of the wines we have made over the years have suffered from any of these interventions in ways that I can detect. Nonetheless, I have been moving toward a less-is-more style of winemaking. Some of the issues, like potential alcohol and pH, I have been able to control more effectively by careful management of when we pick the grapes and what sections of the vineyard we pick from. As for yeast nutrients, I have found over time that our grapes generally provide adequate nutrition for complete, healthy fermentations without supplements. What remains are commercial yeast and bacteria cultures added to ferment the wine and sulfites to inhibit unwanted spoilage microbes that might otherwise ruin the wine. While I have in some years experimented with natural fermentations that rely on yeasts and bacteria cultures that naturally inhabit our grapes and winery, I have not yet reached the point at which I am ready to bet an entire vintage on these unpredictable fermentations. All in good time.

14

LOOKING FORWARD

$\mathbf{M}$ore than 165 years after the first oidium epidemic in France, powdery mildew remains the most common and expensive pest of wine grape vineyards around the world. Each year, winegrowers incur significant costs purchasing and applying fungicides for mildew control. In addition to the direct costs of the fungicides, there are additional labor, fuel, and equipment costs to apply the fungicides. For example, in California alone, there were nearly 45 million pounds of fungicide applied to grapes in 2002, at a total cost of about $123 million, including product and application costs.

The financial costs and risks associated with mildew control are sufficient to drive continuing innovations in equipment and fungicide materials, mostly aimed at better mildew control and reduced fungicide usage. For example, the modern electrostatic sprayer allows growers to achieve better coverage and protection with a significantly smaller volume of fungicide than with traditional air-blast sprayers. Also, the development of systemic fungicides enabled growers to achieve better mildew control with fewer spray applications than had ever been possible with contact fungicides such as sulfur, copper, or various other organic fungicide materials. Largely due to resistance

issues that reduce the effectiveness of the systemic fungicides with repeated usage, the development of new and different systemic fungicides is likely to continue into the foreseeable future. However, most
of these innovations will likely be driven by shortcomings in the existing products and are not likely to fundamentally alter the mildew-
control landscape.

Though the costs of mildew control for the grower are significant,
the broader environmental costs associated with mildew control are
more likely to drive the most important future breakthroughs in mildew control. The handling and application of fungicides may expose
farmworkers to contact with potentially toxic or cancer-causing materials. Fungicide applications can be detrimental to certain wildlife
species in or around the vineyard, such as beneficial insects or aquatic
life in nearby streams and lakes. Finally, there are unknown risks associated with fungicide residues in the wines we consume.

Most fungicides used in grape production are generally believed to
be safe for humans and most wildlife when used according to the instructions. That said, you would be hard-pressed to find many who
would disagree that the use of fewer pesticides in wine (or food) production would be better for our environment, in general, and for the
people who consume wine or other agricultural products treated with
pesticides. While the risks of fungicide usage to control powdery mildew may be acceptable, there are paths the wine industry can follow
to reduce these risks, primarily by reducing the amount of fungicides
used and reducing the potential toxicity (to other than mildew fungi)
of the fungicides that are used.

An important trend in winegrowing to reduce the potential toxicity
of fungicides used in wine grape production is the reduced use of sulfur and copper as fungicides to treat powdery mildew and other fungal
diseases such as downy mildew. Though both sulfur and copper are
naturally occurring elements in the earth, and both are approved for
organic production in the United States and Europe, each poses dan-
gers that warrant minimizing their use for control of powdery mildew.

In the case of sulfur, skin contact or inhalation by vineyard workers can cause serious health problems, with exposure primarily from mixing or applying the material as a fungicide. Sulfur is also very hard on many vineyard insect species, both pests and important beneficials. Copper, a heavy metal, can also cause severe health problems for people and many other species with repeated or excessive exposure. Though not particularly important for powdery mildew control, it is widely used to control downy mildew. Recent regulatory changes in the European Union have reduced the allowable annual usage of copper for mildew control.

Sulfur use has been declining for several decades, initially as many growers switched to systemic fungicides in the 1980s and 1990s and more recently as organic growers have moved away from sulfur in favor of other organic alternatives. However, sulfur remains the single most widely used fungicide for powdery mildew control, primarily because it is so effective. Sulfur also presents little or no risk of fungicide residue contamination in the final product, apart from the potential fermentation issues discussed earlier.

The growth in organic wine production is another important avenue for reducing fungicide toxicity in vineyards. This was an important factor in my own conversion to organic farming, eschewing the use of systemic fungicides and as well as sulfur in the process. My philosophy is quite simple and more or less boils down to one simple question one can ask about any pesticide: Would I spray this on crops that I was raising to feed my family? For me, systemic fungicides could not pass that simple test. I suspect that if every grower looked at his or her vineyard through that simple lens, pesticide usage would be quite different than it is.

Unfortunately, organic production remains a tiny current in the greater body of wine production. Certified organic wine grape acreage in California grew 23 percent from 2008 to 2011 according to California Certified Organic Farmers statistics, but this still represents just over 2 percent of total vineyard acreage in the state. Organic farming for wine grapes is increasing on a worldwide basis, not just here in the United States, with European countries such as Italy, France, Germany, Spain, and Austria leading the way. According to the Research Institute of Organic Agriculture, an independent European nonprofit research organization, organic grape production grew 149 percent worldwide from 2004 to 2010, representing just under 3 percent of total production. Continued growth in organic (and biodynamic) winegrowing will play an important role in reducing pesticide residues in the wines we consume, as well as reducing the use of synthetic chemicals. Given the disproportionate share that grape fungicides represent in overall pesticide usage, this remains an important trend.

Apart from the growth of organic and biodynamic production, one of the most important and promising prospects for reducing pesticide use in wine grape vineyards is research into the development of disease-resistant grape varieties. The biggest challenge, of course, will be to develop disease-resistant variants of the popular wine grapes we all know and love today, such as Zinfandel, Cabernet Sauvignon, and Pinot Noir. The big question: Is this really possible?

In the aftermath of the phylloxera and mildew epidemics of the nineteenth century, there was a great deal of interest in the United States and Europe in the development of hybrid grapevines that would blend the desirable disease resistance (or cold hardiness) of certain American grape species with the wine quality of the European *Vitis vinifera* grapes. However, none of these hybrids successfully transferred the disease-resistance characteristics without also bringing along the undesirable sensory qualities that made their progenitors undesirable as wine grapes in the first place. Also, crossbreeding grape varietals was very slow and laborious, playing out over many years, if not decades, as vines had to be grown from seed or cuttings to maturity before assessing disease resistance and wine quality for each iteration in the hybridization process.

It has long been known that several species of grapes other than *Vitis vinifera* have some level of resistance to common grapevine diseases, including powdery mildew. Most notably, these were native North American grape varietals that coevolved with the powdery mildew fungus *Erysiphe necator* over millions of years in what is now the eastern United States. However, the exact nature or biological basis for this resistance remained a mystery for decades. While not yet fully understood, research has uncovered the basic genetic mechanisms underlying powdery mildew resistance in certain grape varietals.

One such discovery is the locus of the powdery mildew resistance genes in the native American grape known as *Muscadinia rotundifolia*. First identified by researchers at the French National Institute for Agricultural Research, this Muscadinia powdery mildew resistance gene has been given the name Run1, which is a cute little acronym for the first gene found for resistance to *Uncinula necator*.[1] It appears that the Run1 gene causes grapevine tissue cells to die at the site of powdery mildew infection, thereby cutting off the nutrient supply and preventing successful establishment of an infection.

Since the discovery of the Run1 gene, this *Muscadinia rotundifolia* gene has been transferred into other mildew-susceptible grape varieties both by crossbreeding and by genetic transformation. Laboratory and field studies have shown that transference of the Run1 gene introduces substantial resistance to powdery mildew infection into grape varieties previously known to be highly susceptible. Though an exciting and promising finding, it also comes with a few cautions. One of the most important cautions brings us back to evolution and the potential for adaptations to skirt this newfound resistance.

While we might think of grape powdery mildew as a single, worldwide disease or pest of grapevines, it turns out that there are numerous variants or races of grape powdery mildew. It seems that all of the powdery mildew found in the vineyards around the world, outside vineyards in the ancestral home of powdery mildew in the eastern United States, are of a single race and share a common genetic lineage.

---

[1]*Uncinula necator* was the botanical name for the powdery mildew fungus prior to the now widespread use of *Erysiphe necator* as the formal name.

In contrast, there are numerous variants or races of grape powdery mildew in its native region of the eastern United States, with considerable genetic diversity. This genetic diversity means that powdery-mildew-resistant grapevines such as *Muscadinia rotundifolia* are more resistant to some races of powdery mildew than others. Similarly, the effectiveness of individual resistance genes, such as Run1, also varies according to the genetics of the powdery mildew the vine encounters.

The cautionary message here is that the Run1 gene may be part of the answer but by itself will not be sufficient to provide durable, lasting resistance.

Since the initial identification of the Run1 gene, additional gene sequences have been discovered that also play a role in mildew resistance in *Muscadinia rotundifolia* and in *Vitis romanetii*, an East Asian grape species. If these additional gene sequences act by different mechanisms than Run1, then the potential exists to create a more complex resistance mechanism that would be more durable and more difficult for the mildew fungi to overcome.

Another interesting development with very similar implications to the *Muscadinia rotundifolia* resistance genes comes from a most surprising source: *Vitis vinifera* itself. Several obscure central Asian *Vitis vinifera* cultivars have demonstrated considerable resistance to powdery mildew infection and have been used by central Asian breeders to develop mildew-resistant hybrids. In 2008, a Hungarian-led research team identified the genetic locus for this resistance in a *Vitis vinifera* table grape known in its native Uzbekistan as *Kishmish vatkana*, which they have called the Ren1 gene. As with the Run1 gene, when introduced into mildew-susceptible *Vitis vinifera,* it provides substantial resistance to mildew infection. Similar resistance has been found in other central Asian and Chinese *Vitis vinifera* varieties, presenting the opportunity to identify multiple, diverse native vinifera mechanisms for mildew resistance.

The existence of powdery mildew resistance in *Vitis vinifera* grape varieties is both surprising and puzzling. In the absence of any opportunity to develop this resistance in a coevolutionary environment, it would appear that some other more generalized mechanism for responding to fungal pathogens or other abiotic stresses must be underlying the resistance. Whatever the origin, crossbreeding these resistance traits into high-quality wine grapes would be easier or less problematic than would be the case with the non-vinifera sources, such as *Muscadinia rotundifolia* and *Vitis romanetii*. Inevitably, crossbreeding these traits from non-vinifera sources brings along some of the undesirable wine-quality traits that make these grape species unsuitable for wine in the first place.

Remarkable advances in genetic sequencing and identification have dramatically altered the prospects for breeding specific traits into perennial crops like grapes. Conventional breeding protocols required years to propagate and raise grapevines from seeds to sufficient maturity to test for inheritance of desired traits such as powdery mildew resistance. Today, using sophisticated genetic marker identification technology, scientists can analyze plant tissue from newly germinated seedlings to identify the presence or absence of the genetic markers for powdery mildew resistance. It still takes at least two years to get from a seed to a flowering vine for cross-pollination, but the time required to confirm the presence or absence of key inherited traits in the cross-pollinated offspring can be reduced from years to months.

So, the good news is that natural, biological mechanisms have been identified that hold out the prospect of growing high-quality wine grapes with dramatically reduced fungicide usage. While it is unlikely that these various resistance genes will enable winegrowers to eliminate fungicides altogether, only a small fraction of the fungicides used today will be needed.

Sounds too good to be true, doesn't it? Well, unfortunately it may be just that. Not because it cannot be done, but for reasons of market, political, or cultural resistance. Let's start with crossbreeding resistance genes into popular wine grapes such as Cabernet Sauvignon or Pinot Noir. Due to the practical realities and genetics of cross-pollination, the resulting mildew-resistant wine grapes would no longer be Cabernet Sauvignon, Pinot Noir, or any other currently known wine grape varietal. Even though you could use crossbreeding techniques to produce a wine varietal that looked and tasted like Cabernet Sauvignon, with a genetic profile that was nearly 100 percent *Vitis vinifera*, it could never be more than 50 percent Cabernet Sauvignon.

All of today's popular wine grape varietals were produced originally through just this type of crossbreeding, whether intentionally through human intervention or accidentally by pollen carried on the wind between adjacent vines of different varietals. Cabernet Sauvignon itself is the offspring of the cross-pollination of Cabernet Franc and Sauvignon Blanc.

But the realities of the wine market today make the introduction of new wine varietals extremely difficult. In some markets, most notably France, governmental regulations define precisely what types of grapes can be grown in a specific region and may be used in a defined type of wine such as Bordeaux. In these cases, centuries of tradition make any changes in the grape content of such wines extremely difficult. On the broader world stage, wine consumers and critics have long-standing and strongly held convictions on the merits (and deficiencies) of particular types of wine as defined by the grape varietals from which they are made. Cabernet Sauvignon is king. A similar or even nearly identical wine, made from any heretofore unheard-of (mildew-resistant) grape varietal, would face incredible market resistance. As a result, initial use of future mildew-resistant cultivars capable of producing *Vitis vinifera*–quality wines will likely be limited to blending in generic, nonvarietal wines and assorted other markets where climate and disease pressure otherwise prevent the successful growing of traditional wine grape varieties.

An alternative, with an even steeper hill to climb than for the acceptance of new hybrid grape varietals, is the use of genetic transformation to introduce specific resistance genes into popular wine grape varietals. Arguably, a Cabernet Sauvignon vine into which a small number of mildew resistance genes have been inserted could still be called Cabernet Sauvignon. The benefits are potentially considerable. Wine quality and character could be maintained to meet existing sensory profiles, while nearly or completely eliminating fungicide usage in the vineyard and residues in the resulting wines. Unlike many existing genetically modified crops, which are modified specifically to tolerate pesticides, in this case the vines would be modified to reduce or eliminate pesticides. However, unknown risks associated with genetically modified foods have created huge barriers to market acceptance. While certain commodity crops such as corn and soybeans are dominated by genetically modified strains, there is widespread resistance to genetically modified foods in the United States, Europe, and other major wine-producing and -consuming countries. Regardless of how effective transgenic grapevines might be in reducing pesticide usage, it seems extremely unlikely that there will be any commercial acceptance of such vines for a very long time.

15

TRUCE

It seems quite appropriate that, as I sat down to write this final chapter, the first signs of bloom began to appear in our vineyard—appropriate, as the onset of bloom coincides with the period of peak mildew risk and so begins another season of high alert in the vineyard. When we set out to become winegrowers, nearly fifteen years ago at this point, we never imagined how prominent a role death and disease would play in our lives as farmers. Vines, pets, and livestock make up a big part of life on our farm, all intertwined to make life interesting and challenging. So it is, in the annual cycle of life in the vineyard, that we once again return to the always interesting and challenging battle to subdue the forces of powdery mildew in its annual quest to consume our grapes before we can turn them to wine.

I find the annual cycle of life in the vineyard fascinating, with a rhythm all its own. I liken it somewhat to baseball, with the buildup of anticipation that starts when pitchers and catchers report to spring training in February, the excitement of Opening Day as April unfolds, and the glory of the World Series in October. Life in the vineyard follows a similar cycle. As the days begin to lengthen in late February and early March, our thoughts turn to the coming spring with anxious anticipation of the new season. That anticipation turns to excitement as the vines burst forth with new life at bud break near the end of March, culminating in the joy of harvest and fermentation months later in September and October.

It is what lies between these anchor points, bud break in the spring and harvest in the fall, that makes winegrowing interesting and challenging. Every year is different, and the nuances of each season accumulate to define the season and the vintage, placing a unique stamp on each. At the same time, there is a rhythm that transcends these seasonal perturbations. This rhythm follows the life cycle of the vine, from bud break to bloom and fruit set, to the slow growth of the berries that culminates in the onset of ripening at veraison, to the home stretch that follows and then harvest.

There is beauty and wonderment in this annual rhythm, and it is the source of the joy we get both as winegrowers and as wine lovers. But it does not come easy. The beauty of nature is rivaled only by its cruelty, and this is as true in our vineyards as it is on the plains of Africa. The objects of our affection, those grapes that would become wine if given the chance, navigate a perilous course from bud break to harvest each season. It starts with the threat of frost just after bud break, lurking in the cold, clear mornings of late March and early April. Just a few hours below freezing, and the promise of the new season can be decimated before it really begins.

More often than not, though, we push through these cold mornings, and the vines continue their reach up through the trellis toward the sun. Ah, but these tender young shoots are so tasty and irresistible, drawing deer from far and wide in search of a grape leaf feast. In midseason, when the vines are big and bushy, a little deer feeding is no big deal, but in those first few weeks of the season, a small herd of deer can wreak havoc in a single night.

As the season progresses, there may also be thrips, leafhoppers, caterpillars, mites, and other bugs that also seek a steady diet of grape leaves, unknowingly decimating the leafy canopy we rely on to ripen our grapes. Of course, there is always the villain of our story, powdery mildew, and in many vineyards, its cousins, such as downy mildew, black rot, botrytis, and the like. Should we survive all of these and get our nice clean grapes into veraison on the homestretch to harvest, all those lovers of sweet juicy fruit come from far and wide. Bees, birds, raccoons, foxes, squirrels, and who knows what other assorted critters descend on our fields to pillage the fruit of our vineyards.

Given these ravages of nature, I sometimes wonder how we ever make it through to harvest. But we do. It's never easy, and if it's not one thing, it's another that threatens to derail our wine ambitions. But even these trials and tribulations are part of what makes winegrowing the fulfilling work and lifestyle that it is. The natural beauty of the vineyards and the rewards of the wines that follow are what make it all worthwhile.

In the storybook version, wine would be the idyllic end point of a natural, renewable journey each year from bud break to bottle. Winegrowing would be tending to the needs of the vines, nurturing the soil and the vines to achieve a natural balance of fruit and foliage, with minimal intervention or extraneous inputs. This is the essence of the so-called natural wine movement and a worthy aspiration for any winegrower.

In truth, many growers would not even acknowledge this as a worthy aspiration. Agriculture, they would say, is not the natural state of any farmland, and intervention and extraneous inputs are the stock-in-trade of farmers everywhere. While true, there are consequences for these interventions and inputs that extend beyond the annual bounty of harvest. Too little intervention and too few inputs can leave the soil depleted or allow pests and diseases to thrive. On the other hand, too much of either can despoil a beautiful habitat and pollute our ground, air, and water.

How this aspiration yields to reality plays out differently each year for each winegrower. Every season we all battle to save our wine from the ravages of nature. I cast oidium for the lead part in this story because of the central role powdery mildew plays in this battle. By one estimate, if left untended powdery mildew would destroy 97 percent of the wine grapes in California in many, if not most, seasons.

Many critics blame agriculture itself for these ravages of nature, pointing to a monoculture of vineyards and the natural problems that arise in such environments. Unfortunately, these monocultures do exist and do engender their own problems. There are too many vineyards where nothing but grapevines grow for hundreds and hundreds of acres. But it would be unfair to paint all winegrowers with this brush, as many growers strive to maintain diverse habitats, both within and surrounding their vineyards. This is the mind-set of most organic and biodynamic winegrowers and increasingly of those in the sustainable winegrowing movement as well.

Unfortunately, the threat of pests, disease, and predation is the natural state of nearly any organism in any habitat. Grapevines are no different. What does differ is how we as farmers and winegrowers respond to these threats. In large part, this book has been about these threats, how winegrowers in general deal with them, and more specifically our own philosophy and experiences battling these ravages of nature.

Often when reflecting on this and other challenges that make farming difficult, I return to consider the most common refrain we encounter when entertaining guests, mostly on their first visit to our vineyard. "You are living the dream." I cannot begin to tell you how many times we have heard this refrain, and I certainly would not argue against it. It is a dream come true; it's just that the original dream seems to have left out a lot of the less glamorous details!

I find it ironic that this dream, hatched by two well-educated and well-paid professionals with postgraduate degrees, would include many aspects of a lifestyle that centuries of our forebears struggled to leave behind. Hard work in the hot sun for little or no pay. When put like that, it hardly seems glamorous. But it is a healthy and rewarding lifestyle. Not once since we got here have we pined for the good old days in the air-conditioned office, behind a desk or in a meeting.

There is an almost visceral satisfaction that comes anew each year from producing wine with our own hands from the land and vines we work so hard to nurture. That satisfaction is reinforced continually as we consume the fruits of our labor each night with dinner and share these wines with friends and family, whatever the occasion.

Winegrowing also brought about a fundamental change in our relationship to wine. Lost were some of the romance and mystery, but those were replaced with a deeper, more intimate appreciation. Years ago, before we embarked on this vineyard odyssey, I was an enthusiastic wine consumer and collector. Never really a collector of trophy wines, I sought out interesting wines from around the world, seeking value, variety, and occasionally just novelty. I purchased most of my wine on special buying trips to major wine outlets or to the California Wine Country, sometimes traveling a hundred miles or more to reach specialty wine shops with unusually large or diverse selections. I devoured publications such as *Wine Spectator* and kept lists of wines that I would seek out on these trips based on the tasting notes in the wine reviews. I cellared wines with a system organized around the projected date or year of consumption, with special wines allocated for uncorking ten years or more after purchase. This was a period of discovery, lasting nearly two decades, and it created the foundation and fostered the passion for what we are doing today.

This passion, or some variant thereof, is the engine that drives the fine-wine market and Wine Country tourism as well. Wine appreciation becomes a complex mix of sensory experiences (i.e., drinking and tasting), combined with other psychological influences such as the wine's price, the reputation of its label, producer, or place of origin, and the august opinions of wine reviewers such as *Wine Spectator* or Robert Parker. For the purposes of this discussion I would describe this passion as from the outside looking in.

It is precisely that perspective that changed for me as my role changed from consumer and collector to grower and producer. Now I find myself on the inside looking out. I don't mean to imply that this perspective is better or more insightful, only that it is fundamentally different. It is different in ways that have transformed my personal relationship with and passion for wine. Now I cannot drink or taste wine without thinking about where, how, and by whom the wine was grown and produced. How is it similar to or different from what I do? Do I approve or disapprove of their various methods and practices? I still read *Wine Spectator,* but now I read the reviews, looking for the wines of my friends and neighbors and wines that I might consider competitors of the wines produced from our vineyard.

I no longer accumulate lists of wines to buy and—no surprise—buy far less wine than I use to. When I do buy wine, it is often just to reach out and try wines from different parts of the world. Generally I do this without a plan or purpose but driven by the need to experience wines that are different from the local wines we drink every day. In our business it is easy to develop what we call a house palate, where you drink so much of your own wine that you begin to develop a very narrow view of what good wine is supposed to taste like. Most of the wine we drink now is either grown or made by us or by our friends and neighbors. Despite the diversity of wine grapes and styles in Sonoma, compared to the broader diversity of wines around the world, one still develops a rather narrow palate from such a local focus.

Whatever your perspective on or relationship with wine, it can only be enhanced peering more intently from the inside out. You need not be a grower or producer to do this. The more you know about the where, the how, and the by whom of the wines you drink, the more you will be able to appreciate their particular qualities and differences. It will also help you make choices that reward and encourage wines that are grown and produced in certain ways, hopefully in ways that nurture and sustain us as people and our environment.

That's all part of what makes wine so fascinating.

appendix

THE BASIC BIOLOGY OF MILDEW INFECTIONS

To understand the whole picture of how powdery mildew gets started and spreads through a vineyard, it helps to look at the various ways that mildew can infect a vineyard. Basically there are three ways for a powdery mildew infection to get started in a vineyard each season:

1. Ascospores: These spores are released during warm winter and spring rains from mildew "seeds" deposited in the vineyard the previous summer and fall. These seeds are produced by mildew infections during the summer and fall, after which they hibernate over the winter in the crevices of the vine bark.
2. Flag shoots: These are mildew infections that penetrate newly formed buds in the spring and then lie dormant over the following winter under a protective scale covering the nascent buds. They emerge as live mildew infections when the infected buds open to provide the new season's growth the following spring.
3. Conidia: These are reproductive spores, similar to ascospores, that are released by mildew infections during the growing sea-

son, spreading infections from vine to vine and from vineyard to vineyard.

The first two routes to infection, ascospores and flag shoots, provide the means by which mildew infections persist in a vineyard from season to season. Consequently, these represent the primary mechanisms for the initial infections in any given growing season. Conidia, on the other hand, provide the mechanism by which an actively growing mildew infection actually spreads from vine to vine and even from vineyard to vineyard. Conidia can also be the source of an initial infection when they are blown by the wind from an infected vineyard to a nearby vineyard.

Since powdery mildew requires tender, green, growing tissues for nourishment and reproduction, it is quite difficult for live mildew colonies to survive the winter in the vineyard. Other than the unique conditions that produce flag shoots, winter temperatures and dormant grapevines are not a suitable environment for sustaining mildew over the winter. So each spring at bud break, when the first tender growth emerges from the carefully pruned vines, most vineyards start anew, free of mildew infection. But dormant mildew spores lie hidden, ready to infect the tender new leaves and shoots at the first opportunity.

Ascospores

The primary source for powdery mildew infections in many, if not most, vineyards comes from ascospores released by tiny mildew bodies called cleistothecia. The cleistothecia are like microscopic seedpods, as they are small protective enclosures about one hundred micrometers in diameter, which is about the width of a human hair. Each seedpod contains a few dozen spores. They are produced by mildew infections growing on the surfaces of leaves, shoots, and grapes clusters during the late summer and early fall. A relatively minor infection on a single vine can produce thousands upon thousands of these tiny seedpods, which are later washed off of these surfaces during the fall and winter rains, falling into the cracks and crevices of the vine bark. It is in the bark of the vine trunks and cordons where they spend the winter, lying in ambush to infect new vine growth the following spring. When spring rains wet the cleistothecia, they open up and release their spores into the air. Infections occur when these ascospores land on green, growing grapevine tissue and germinate. If the spores land anywhere else, such as on the ground or the lignified woody trunk of the grapevines, they will expire without germinating.

Cleistothecia also fall to the ground when they are washed off with the runoff from winter rains, as well as when the leaves fall and pruned wood is dropped on the vineyard floor. While some of these may survive the winter, the viability of cleistothecia on the ground, on fallen leaves, or on pruning debris is much lower and presents a much smaller infection risk than those overwintering on the vine itself.

So in any vineyard where there was some mildew infection the prior season, even tiny infections that went unnoticed, there are probably cleistothecia lurking on the vines the following spring. The likelihood and severity of any infection caused by spores released from these cleistothecia are somewhat a function of how severe the infection was in the prior season, which determines how many cleistothecia were produced in the first place. Also, rainy weather prior to leaf drop in the fall is required to wash the cleistothecia onto the vine bark. Finally, rainfall in the following spring is required to release ascospores and start the cycle of infection.

For example, even following a relatively severe infection in the summer, if the fall season is very dry without sufficient rainfall to wash the cleistothecia off the leaves and canes onto the vine bark before the leaves fall off, there may be relatively few viable cleistothecia to release ascospores the following spring. Those cleistothecia that do make it onto the bark require very specific conditions the following spring to release their spores and cause a mildew infection. First off, the cleistothecia require a sustained wetting period, generally in the form of at least a tenth of an inch of rain that wets the cleistothecia for a period of thirteen hours. Second, ambient temperatures during and after the rainfall must be within a certain range, with a minimum temperature of 50°F and a maximum of 80°F. Third, there must be susceptible green vine tissue, such as young leaves, shoots, flowers, or berries, on which to land and germinate.

Ascospores are produced in the cleistothecia in small sacs of four to eight spores each. These sacs are called asci (singular: ascus). When the spores are released, being only a few micrometers in diameter, they float in the air and are carried by wind or other air currents. Infection occurs when the spores land and germinate on green, living vine tissue. Grape powdery mildew will only germinate and grow on grapevines. No other plants are suitable hosts.

When the ascospores germinate, they produce tiny hairlike filaments (hyphae) on the surface of the grapevine tissue. These microscopic filaments send roots into the vine tissue, through which the mildew feeds on nutrients in the vine tissue. These roots are called the haustoria. Once the ascospore germinates and sends its roots into the tender vine tissue, the hyphae begin to multiply on the surface of the leaves and grapes, forming an interconnected web (mycelia) that appears as the characteristic white, powdery coating that gives the mildew its name. Penetration of the vine tissue by the haustoria and growth of the mycelia on the surface eventually kill the vine tissue in the immediate area of the infection, which causes the scarring and discoloration typically seen on infected leaves, shoots, and berries.

Ascospore release is the most important mechanism for the initial infection of vineyards each season in California and in many other winegrowing regions of the world. While very specific conditions are required for infection, these conditions tend to be quite common. Nonetheless, if ascospores are released during winter or spring rains before bud break, no infection will occur, as there are no green growing tissues to colonize. However, once bud break happens and tender green tissue is present, any ascospore release must be considered to be an infection event. This means that almost any time it rains after bud break, new infections are likely unless protective measures are taken. This generally remains true until midsummer, by which time all of the cleistothecia have either released their ascospores or are simply no longer viable.

Flag Shoots

Most mildew infections in the spring occur initially on the bottoms of the basal leaves. Basal leaves are the first few leaves produced on new shoots each season, so called because of their location at the base on the shoot. Basal leaves are the closest leaves to the cordon and vine trunk bark where cleistothecia hide and provide a broad flat surface for the ascospores to land on. In some instances, infections on basal leaves or adjacent shoot tissues lead to the formation of flag shoots.

Flag shoots occur when a mildew infection penetrates into the newly formed buds that will become the basis for the next season's growth. On grapevines, buds are formed at the junctions where the leaf stems connect to the shoots on which they grow. These buds, at the base of the leaf stems, will be the buds from which next year's shoots grow. If a mildew infection occurs on the outer surface of these buds at a critical stage of development, the mildew is able to penetrate under the protective scale covering the bud and into the bud itself. This allows the mildew to actually colonize the bud, living under a protective scale that covers the bud. The bud scale provides a protective environment for the mildew to survive the upcoming winter and emerge with the new shoot growth the following spring.

The most critical period for infections that lead to flag shoots occurs when the new shoots have from three to six new leaves in the spring. At this time, the young buds forming on the shoots are susceptible to penetration by mildew infection. Beyond this point, the protective scale on the buds hardens to provide a shield against penetration by mildew infections. Mildew infections that colonize these young buds lie dormant under the protective scale until bud break the following spring.

Flag shoots are more common in some viticultural regions than others. They tend to favor hot or very warm regions and are less common in regions with cold winters. When flag shoots occur they are recognized by the characteristic curling of the young grape leaves and stunted growth at the shoot tip. The shoot will be stunted and the leaves malformed, and the entire shoot will be covered by the powdery white mycelia of the mildew. At this point, the flag shoots present a significant source of infection for mildew to spread to other vines and neighboring vineyards. Infections on flag shoots spread when spores are released from reproductive structures in the mycelia. Spores produced in the mycelia are called conidia, which are carried by winds and air currents. When they land on green, growing grapevine tissue, they germinate to create new infections. Conidia are the reproductive spores produced by active mildew colonies, which may result from the emergence of flag shoots in the spring or the germination of ascospores as described above.

Conidia

The powdery mildew fungus has two primary reproductive mechanisms, both of which involve the release of spores into the air. The ascospores, as described above, exist primarily as a means for mildew to propagate from one growing season to the next, overwintering from one growing season to the next in the bark crevices of the vines. Once the growing season begins, conidia sporulation becomes the primary reproductive mechanism and the engine that powers serious mildew epidemics.

Conidia are produced in special reproductive structures in the mildew hyphae, which are the filament-like structures that grow on the leaf or berry surface after a spore germinates. As individual hyphae mature, they produce anywhere from one to ten conidia spores in a chain at the tip of the filaments.

Under ideal conditions, newly germinated spores can mature, grow secondary hyphae, and release conidia spores in as few as five days. As they have the ability to reproduce up to tenfold every five days, you can see how an infection can rapidly spin out of control. This is the primary mechanism for the spread of mildew during the growing season, and it is dependent on sunlight and temperature more than moisture.

Powdery mildew spores can germinate at temperatures ranging from 33°F to 92°F. After germination, the optimum temperature for growth and reproduction is about 75°F. However, maximal germination, growth, and reproduction occur in the range of about 70°F to 85°F. Within this favorable range of temperatures, the time between spore germination and reproduction (release) of new spores is only five to seven days. Mildew risk is greatest in growing areas or seasons where daytime and nighttime temperatures are near or within this range.

Fortunately very high and very low temperatures are detrimental to mildew growth and reproduction. Mildew spores and colonies begin to die and reproduction drops significantly when temperatures exceed 96°F for several hours. Mildew infections can be destroyed completely if temperatures rise above 105°F or fall below 45°F for only a few hours.

In addition to temperature, exposure to direct sunlight plays a significant role in mildew growth and conidia reproduction. Direct sunlight slows mildew growth and reproduction. Furthermore, direct sunlight interacts with temperature such that the inhibitory effects of sunlight exposure increase with increasing temperatures. Even within the optimum temperature range, several hours of direct sunlight exposure per day can as much as double the number of days required for reproduction.

Although powdery mildew can thrive in low-humidity environments, mildew severity does increase with increasing humidity, with an optimum of about 85 percent relative humidity. Hence fog and residual moisture from rainfall, or even overhead sprinkler irrigation, can play a significant role in reproduction. In many areas, such as

California's coastal winegrowing regions, foggy weather brings the double threat of low sunlight and high humidity. Similarly, heavy vine canopies with lots of leaves to provide shade and retain humidity from fog and rainfall also greatly increase the risks of mildew epidemics.

THE SYMPTOMS AND CONSEQUENCES OF MILDEW INFECTIONS

As noted earlier, most vineyards wake up in the spring free of mildew infection. That is different from being free of mildew, as many vineyards contain dormant mildew spores hiding in the bark of the grapevine trunks and cordons. There are likely very few vineyards in any of the major winegrowing regions of the world that are completely free of any mildew spores or spore-producing mycelia. The challenge for winegrowers every year is to prevent this ever-present threat from turning into actual thriving, reproducing mildew colonies. Once an infection occurs, the stage is set for potential disaster.

Infections are most likely to occur and grow on the bottoms of leaves or on grape clusters and other grapevine tissues that are shaded by the vine canopy. Spore germination and mildew growth and reproduction are all inhibited by direct exposure to sunlight. Early in the growing season, shaded vine tissue that is hospitable to mildew infection is often limited to the undersides of the leaves, as most of the new vine growth is exposed to direct sunlight for much of the day. However, as the vine grows and the canopy gets larger, more and more of the season's new growth is shaded from direct sunlight by the heavy foliage of the canopy and, hence, more hospitable to mildew germination, growth, and reproduction.

The first signs of infection are often yellowish splotches on the top and bottom surfaces of leaves. Upon closer inspection, the underside of these yellow spots will be populated with the powdery white web of the mildew mycelia. If left unchecked, the mildew will continue to grow on the leaf surface, causing the entire leaf to yellow and eventually fall off. In the meantime, of course, the infected leaf is a source of infection for nearby leaves, shoots, grapes, and adjacent vines as well. Under ideal conditions, the mildew can grow very rapidly, colonizing much of the leaf's surface before notable damage is visible. When this occurs, the leaf will usually be covered extensively by the characteristic powdery white mycelia.

Of course, infections are not limited to the leaf surfaces. The main branches or canes of new shoots infected early in the season may exhibit the same powdery white mycelia. Often infections will appear as brown spots or scars on the surfaces of the canes.

Mildew-infected leaves and shoots can adversely affect the growth of the shoot as well as the photosynthesis needed for the vine to grow properly, not to mention to produce and ripen fruit. Fortunately, grapevines tend to be robust, vigorous plants. Small amounts of foliar damage are easily overcome as new leaves and secondary shoots are produced almost continuously from bud break to the onset of ripening (veraison).

The real damage from mildew-infected vines occurs when the fruit itself is infected. The risk of fruit damage starts before flowering, when the flower clusters are most susceptible to infection. Infection of the flowers can be especially troublesome. Severe infection can disrupt flowering and pollination, with a resulting reduction in yields. Infection of the flower cluster, even if pollination is not disrupted, puts an infection in the core of the newly formed grape cluster during the critical stages of flowering and fruit set. The likelihood of fruit infection at this point is very high.

Mildew-infected fruit is not a pretty sight. The first signs of fruit infection appear as a powdery white coating on the berry surfaces. Closer inspection will usually find infection on the rachis, or stems of the grape cluster, as well. As the infection progresses, the characteristic white coating of the berries is accompanied by scarring of the berry skins as well. This scarring of the berry surface will cause problems later in the season as the berry grows and enlarges. The scar tissue is unable to grow with the berry and ultimately leads to cracking of the skin, exposing the grape pulp and seeds to desiccation and secondary infection.

If treated promptly, early-season infections can be eradicated with minimal fruit damage. The berries are most susceptible to mildew infection in the period from just before bloom until about thirty days after flowering. Preventing mildew infection during this period is the most important factor for ensuring mildew-free fruit later in the season. Infections that simmer undetected during this period can flare up later in the season, sometimes with catastrophic results. Infected grape clusters become covered in a whitish-gray mold and often split open as the scarred skin of the berry is unable to grow with the berry as it enlarges during ripening. If the infection is arrested with fungicide treatments, the infected berries may just dry up and shrivel on the vine. Frequently, the mildew damage opens the berries to secondary infections, such as botrytis, which creates a whole new disease cycle.

selected References

While this book is meant to be entirely factual, it turns out to be nearly equal parts memoir and textbook. Actually, I hope it reads much better than your typical textbook, and I do not intend for it to be authoritative in the vein of a textbook or reference book. Nonetheless, a great deal of research went into many of the chapters in the book. Many of the sources are available in pubic libraries, especially the fabulous Sonoma County Wine Library in Healdsburg, California. Of course, no research project today would be complete without access to the Internet. Internet research sources are many, including online reproductions of countless technical and scientific articles and abstracts, but also without equal is Wikipedia, which is an invaluable starting point for so many research threads. For the chapters recounting my personal experiences, I drew on my own notes and memories. For all of the other chapters, I provide below selected references from which I drew most of the factual or technical reference material on which those chapters rely.

THE ORIGIN OF OUR AFFECTION
AND ITS PECULIAR AFFLICTIONS

Ainsworth, Geoffrey Clough. *Introduction to the History of Plant Pathology*. Cambridge: Cambridge University Press, 1981.

Campbell, C. *The Botanist and the Vintner*. Chapel Hill, NC: Algonquin Books, 2004.

Flagg, W. J. *Three Seasons in European Vineyards*. New York: Harper and Brothers, 1869.

Johnson, H. *Vintage: The Story of Wine*. New York: Simon and Schuster, 1989.

Large, E. C. *The Advance of the Fungi*. London: Jonathon Cape, 1940.

McGovern, P. E. *Ancient Wine: The Search for the Origins of Viniculture*. Princeton, NJ: Princeton University Press, 2003.

Mudge, K., et al. "A History of Grafting." *Horticultural Reviews* 35 (2009).

New Jersey State Board of Agriculture. *Annual Report of the State Board of Agriculture*. Volumes 15–16. 1888.

Ordish, G. *The Great Wine Blight*. London: Sidgwick & Jackson, 1972.

Paul, H. *Science, Vine and Wine in Modern France*. Cambridge: Cambridge University Press, 1996.

Terral, J.-F., et al. "Evolution and History of Grapevine (*Vitis vinifera*) under Domestication: New Morphometric Perspectives to Understand Seed Domestication Syndrome and Reveal Origins of Ancient European Cultivars." *Annals of Botany* 105, no. 3 (2009).

Unwin, T. *Wine and the Vine—an Historical Geography of Viticulture and the Wine Trade*. London: Routledge, 1991.

Walker, A. R., et al. "White Grapes Arose through the Mutation of Two Similar and Adjacent Regulatory Genes." *The Plant Journal* 49, no. 5 (2007).

SOME THINGS GET BETTER WITH AGE

Calonnec, A., et al. "A Host-Pathogen Simulation Model: Powdery Mildew of Grapevine." *Plant Pathology* 57 (2008).

Doster, M., and W. Schnathorst. "Comparative Susceptibility of Various Grapevine Cultivars to the Powdery Mildew Fungus *Uncinula necator*." *American Journal of Enology and Viticulture* 36, no. 2 (1985).

Ficke, A., et al. "Effects of Ontogenic Resistance upon Establishment and Growth of *Uncinula necator* on Grape Berries." *Phytopathology* 93 (2003).

———. "Ontogenic Resistance and Plant Disease Management: A Case Study of Grape Powdery Mildew." *Phytopathology* 92 (2002).

Gadoury, D., et al. "Effects of Diffuse Colonization of Grape Berries by *Uncinula necator* on Bunch Rots, Berry Microflora, and Juice and Wine Quality." *Phytopathology* 97 (2007).

———. "Grapevine Powdery Mildew (*Erysiphe necator*): A Fascinating System for the Study of the Biology, Ecology and Epidemiology of an Obligate Biotroph." *Molecular Plant Pathology* 13, no. 1 (2012).

Gessler, C., et al. "*Plasmopara viticola:* A Review of Knowledge on Downy Mildew of Grapevine and Effective Disease Management." *Phytopathologia Mediterranea* 50 (2011).

Kennelly, M., et al. "Seasonal Development of Ontogenic Resistance to Downy Mildew in Grape Berries and Rachises." *Phytopathology* 95 (2005).

Merry, A. "Ontogenic Resistance in Grapevine Leaves to Powdery Mildew." PhD thesis, University of Tasmania, 2011.

Proceedings of the Fifth International Workshop on Grapevine Downy and Powdery Mildew, San Michele all'Adige, Italy, June 2006.

Proceedings of the Sixth International Workshop on Grapevine Downy and Powdery Mildew, Bordeaux, France, July 2010.

Raymond, W., et al. "Powdery Mildew Induces Defense-Oriented Reprogramming of the Transcriptome in a Susceptible but Not in a Resistant Grapevine." *Plant Physiology* 146 (2008).

THE WINEGROWER'S CHALLENGE

Kast, W., and K. Bleyer. "Efficacy of Sprays Applied against Powdery Mildew (*Erysiphe necator*) during a Critical Period for Infections of Clusters of Grapevines (*Vitis vinifera*)." *Journal of Plant Pathology* 93 (2011).

Lybbert, T., and W. Gubler. "California Wine Grape Growers' Use of Powdery Mildew Forecasts." Giannini Foundation of Agricultural Economics, University of California, 2008, http://giannini.ucop.edu/media/are-update/files/articles/v11n4_4.pdf.

Proceedings of the Fifth International Workshop on Grapevine Downy and Powdery Mildew, San Michele all'Adige, Italy, June 2006.

Proceedings of the Sixth International Workshop on Grapevine Downy and Powdery Mildew, Bordeaux, France, July 2010.

Conventional Farming

Hager, Thomas. *The Alchemy of Air: A Jewish Genius, a Doomed Tycoon, and the Scientific Discovery That Fed the World but Fueled the Rise of Hitler.* New York: Crown Publishing, 2009.

USDA National Agricultural Statistics Service. "California Grape Crush Report," 2011, http://www.nass.usda.gov/Statistics_by_State/California/Publications/Grape_Crush/index.asp.

Sustainable Farming

California Sustainable Winegrowing Alliance, http://www.sustainablewinegrow
    ing.org.
Lodi Certified Green, http://www.lodiwine.com/certified-green.
Napa Sustainable Winegrowing Group, http://www.naparcd.org/nswg.html.
Oregon Low Input Viticulture, http://liveinc.org.

Organic Farming

California Certified Organic Farmers, http://www.ccof.org.
USDA National Agricultural Library, "Organic Production," http://afsic.nal.usda.
    gov/organic-production.

Biodynamic Farming

Demeter Association, http://www.demeter-usa.org.
Steiner, R. *The Agricultural Course*. London: Bio-dynamic Agricultural Associ-
    ation, 1958. Accessed online at the Rudolph Steiner Archives, http://
    wn.rsarchive.org/Lectures/Agri1958/Ag1958_index.html.
Unger, C. *What Is Anthroposophy?* Fair Oaks, CA: Rudolf Steiner College Press/
    Saint George Publications, 1981.

THE ART AND SCIENCE OF MILDEW CONTROL

Gubler-Thomas Powdery Mildew Risk Index, http://www.ipm.ucdavis.edu/DIS
    EASE/DATABASE/grapepowderymildew.html.

NATURE STRIKES BACK

Carson, Rachel. *Silent Spring*. New York: Houghton Mifflin, 1962.
Fungicide Resistance Action Committee, http://www.frac.info.
Gisi, U., and H. Sierotzki. "Molecular and Genetic Aspects of Fungicide Resis-
    tance in Plant Pathogens," Fifteenth International Reinhardsbrunn Sym-
    posium, Friedrichroda, Germany, 2007.

Gubler, W., et al. "Occurrence of Resistance in *Uncinula necator* to Triadimefon, Myclobutanil, and Fenarimol in California Grapevines." *Plant Disease* 80 (1996).

Klittich, Carla J. "Milestones in Fungicide Discovery: Chemistry That Changed Agriculture." *Plant Health Progress* 10 (2008).

McKenna, Maryn. *Superbug: The Fatal Menace of MRSA.* New York: Free Press, 2009.

Palumbi, Stephen R. *The Evolution Explosion: How Humans Cause Rapid Evolutionary Change.* New York: W. W. Norton & Co., 2001.

Steinkellner, S., and H. Redl. "Sensitivity of *Uncinula necator* Populations Following DMI-Fungicide Usage in Austrian Vineyards." *Die Bodenkultur* 52, no. 4 (2001).

FROM VINE TO GLASS

Amati, A., et al. "Preliminary Studies on the Effect of *Oidium tuckeri* on the Phenolic Composition of Grapes and Wines." *Vitis* 35, no. 3 (1996).

Asprinio, S. "The Impact of Powdery Mildew on Wine Quality." *Wines & Vines,* May 2004

Caboni, P., and P. Cabras. "Pesticides' Influence on Wine Fermentation." *Advances in Food and Nutrition Research* 59 (2010).

Calonnec, A., et al. "Effects of *Uncinula necator* on the Yield and Quality of Grapes (*Vitis vinifera*) and Wine." *Plant Pathology* 53, no. 4 (2004).

Gadoury, D., et al. "Effects of Diffuse Colonization of Grape Berries by *Uncinula necator* on Bunch Rots, Berry Microflora, and Juice and Wine Quality." *Phytopathology* 97 (2007).

Ortiz, J., et al. "Influence of Fungicide Residues in Wine Quality." InTech. 2010. http://www.intechopen.com/books/fungicides/influence-of-fungicide-residues-in-wine-quality.

Pesticide Action Network Europe. "Message in a Bottle." 2008. http://www.pan-europe.info/Resources/Briefings/Message_in_a_Bottle.pdf.

Provenzano, M., et al. "Copper Contents in Grapes and Wines from a Mediterranean Organic Vineyard." *Food Chemistry* 122 (2010).

Rose, G., et al. "The Fate of Fungicide and Insecticide Residues in Australian Wine Grape by Products Following Field Application." *Food Chemistry* 117 (2009).

Stummer, B., et al. "Effects of Powdery Mildew on the Sensory Properties and Composition of Chardonnay Juice and Wine When Grape Sugar Ripeness

Is Standardized." *Australian Journal of Grape and Wine Research* 11, no. 1 (2005).

Tromp, A., and C. A. de Klerk. "Effect of Copperoxychloride on the Fermentation of Must and on Wine Quality." Tenth Congress of the South African Society for Enology and Viticulture, Cape Town, 1986.

Werner, M., et al. "Impact of Different Application Rates of Wettable Sulphur and Selected Enological Practices on SO2-Levels and Aroma-Active Compounds in Organic Riesling Wine." Sixteenth IFOAM Organic World Congress, Modena, Italy, June 16–20, 2008.

APPENDIX

Austin, C., and W. Wilcox. "Heat and UV Radiation from Sunlight Exposure Inhibit Powdery Mildew." Appellation Cornell. Research Focus 2010-2. Cornell Viticulture and Enology. 2010. http://grapesandwine.cals.cornell.edu/cals/grapesandwine/appellation-cornell/issue-2/upload/Research-Focus-2.pdf.

Cortesi, P., et al. "Distribution and Retention of Cleistothecia of *Uncinula necator* on the Bark of Grapevines." *Plant Disease* 79 (1995).

Hajjeh, H., et al. "Overwintering of *Erysiphe necator Schw.* in Southern Italy." *Journal of Plant Pathology* 90, no. 2 (2008).

Halleen, F., and G. Holz. "An Overview of the Biology, Epidemiology and Control of *Uncinula necator* (Powdery Mildew) on Grapevine with Reference to South Africa." *South African Journal for Enology and Viticulture* 22, no. 2 (2001).

Rossi, V., et al. "Dynamics of Ascospore Maturation and Discharge in *Erysiphe necator,* the Causal Agent of Grape Powdery Mildew." *Phytopathology* 100, no. 12 (2010).

Acknowledgments

$f$irst off, this book would not have been possible without the encouragement and support of many friends and colleagues in the Sonoma winegrowing community over the years. This fabulously open and sharing community fostered our development as winegrowers in many ways that led to much of the knowledge and experiences that I have related in this book. Sonoma is a place not only of great wine but of great people as well. Also, Steve Hall and Robert Biale, of Robert Biale Vineyards, have been invaluable partners in our vineyard odyssey, without whom the journey that I have related in this book would have been much more difficult and less enjoyable.

Though I received much support and encouragement from many friends in the writing of this book, I must thank my longtime friend Mark Altom for his patience and diligence in reviewing every draft of this book. Mark and I have drunk a lot of wine together over the years, and he was the first person I thought of when looking for a wine enthusiast to review the early drafts of this book. I would also like to thank my friend and neighbor Sherry Rahn, both for her encouragement along the way and for the artwork that she created that appears in the pages of this book.

I would also like to thank my editor, Susan McEachern of Rowman & Littlefield Publishers, for seeing the potential in my manuscript and for bringing to fruition the book you are now reading.

Finally, and most importantly, I would like to thank Deb Kiger, my lovely wife and partner in this vineyard odyssey. Deb has been an invaluable partner every step of the way, from the genesis of the idea that became Kiger Family Vineyard in the first place, to the many winegrowing experiences that we have shared over the years, and finally for her support and encouragement in the writing of this book, as well as her patient reviews of the many drafts along the way.

index

about the author

John Kiger is an accomplished winegrower in California's beautiful Sonoma Valley. John lives and works with his wife, Deb, amid the Syrah, Grenache, and Cabernet Sauvignon vines of Kiger Family Vineyard, where they focus on organic and sustainable winegrowing practices. John is active in the Sonoma winegrowing community and has spoken on organic winegrowing practices in various grower forums. He has been an advocate for innovative organic practices, such as using sheep to control weeds and grasses in vineyards, and has been the subject of articles in industry publications such as *Wine and Grape Grower.*

Little about John's upbringing or education would peg him as either a vineyard owner or a farmer. Born and raised in North Carolina, and with a PhD in psychology and a specialty in human-computer interaction, John embarked on a career in high technology that would lead him to California and the high-pressure, fast-paced world of Silicon Valley. After he had spent more than twenty years in software engineering and marketing, his passion for fine wine that had been simmering under the surface burst forth in a life-changing decision to leave it all behind and become a full-time winegrower. He and his wife left their Silicon Valley jobs and house in the suburbs behind, bought land in the wine-rich Sonoma Valley, and built their very own vineyard from scratch. Thirteen years later that experience spawned the expertise and inspiration for John's first book, *A Vineyard Odyssey: The Organic Fight to Save Wine from the Ravages of Nature*.